New Energy

New Energy

Understanding the Crisis and a
Guide to an Alternative Energy System

James Ridgeway
and
Bettina Conner

An Institute for Policy Studies Book

Beacon Press   Boston

Copyright © 1975 by The Institute for
Policy Studies, Washington, D.C.
Beacon Press books are published under the auspices
of the Unitarian Universalist Association
Simultaneous publication in Canada by Saunders of Toronto, Ltd.
All rights reserved
Printed in the United States of America

9 8 7 6 5 4 3 2 1

Library of Congress Cataloging in Publication Data

Ridgeway, James, 1936-
   New energy.
   "An Institute for Policy Studies book."
   Includes index.
   1. Energy policy—United States. 2. Petroleum
industry and trade—United States. 3. Power resources
—United States. I. Conner, Bettina, joint author.
II. Title.
HD9502.U52R53   333.7   74-16669
ISBN 0-8070-0504-5

For Frank Fragale

Acknowledgments

New Energy grew out of our work for *The Encyclopedia of Social Reconstruction,* a major project of the Institute for Policy Studies. We are indebted to the Institute and especially to the members of the Political Economy Program Center who helped work out the plan for an alternative energy system set forth in Part Four.

Much of the historical research was done by James Ridgeway at the University of California, Davis, where he was a Regents Lecturer; and he is especially indebted to the members of the Political Science Department at Davis for their support.

In addition, we wish to thank Pastel Vann, Bethany Weidner, Robb Burlage, and Leonard Rodberg for their help in this work.

Contents

Introduction xi

PART ONE The Roots of the Crisis: A Brief History of the Energy Industry

   1 The Standard Trust 1
   2 The Modern Industry 6
   3 The International Petroleum Cartel 15
   4 World War II 23
   5 Oil and the Cold War 30
   6 Coal 39
   7 Natural Gas 47
   8 Electric Utilities 52

PART TWO The Energy Crisis of 1973–74 69

   9 The New Industry 70
  10 Oil Shortages 89
  11 Feeble Reform 101

PART THREE Resistance 121

  12 The Georgia Power Project 122
  13 Lifeline Service 136
  14 The Fight Against Pacific Gas & Electric 142

PART FOUR **Proposal for an Alternative Energy System** 153

Chapter Notes 175
Appendix 180
Index 221

Introduction

New Energy attempts to place the recent energy crises in historical perspective and offers a plan for an alternative system.

It is our basic contention that one can get a clear understanding of the current crisis only by examining the history of the oil industry and its relationship with the state. Natural resources will soon run out. But we believe the recent energy crises were not caused by any real shortages of fuels but rather that they occurred because of the way in which the industry is organized and functions within the political economy.

The roots of the crisis go back to the early part of the century, when the modern oil industry took shape. At that time the successor companies to the Standard Oil trust formed themselves into vertically integrated operations that controlled oil from wellhead to gas pump. To establish the industry, the major firms sought the help of state and federal governments in providing laws to control the flow of oil. These arrangements, initiated under Coolidge and consolidated during the New Deal, established relationships that now prevail: the companies make policy while the government acts as their passive agent.

After World War I, the Standard Oil Company of New Jersey led the other big U.S. oil companies in seeking an accommodation with the British in the Middle East. Standard feared a loss of world markets should the British obtain long-term concessions in Iraq. Thus, in 1928 Standard and the Anglo-Persian Oil Company, partly owned by the British government, organized a partnership to control the flow of oil in world markets. This was the Iraq Petroleum Corporation, and the beginning of the "brotherhood of oil merchants," or the international petroleum cartel.

During World War II the U.S. government strengthened the cartel by granting lend-lease aid to Saudi Arabia and making available scarce steel to construct a pipeline the companies needed to further organize their markets. Although the Antitrust Division of the Justice Department attacked the cartel in the early 1950s, President Truman overrode the lawyers and took the advice of his Secretary of State, Dean Acheson, who saw the oil companies as important partners in staging diplomatic initiatives during the Cold War. The United States became more deeply involved in Middle Eastern politics, perpetrating the coup that overthrew the Iranian Nationalist Premier, Mohammed Mossadegh. That, in turn, led to reorganization of Iran's oil industry, opening participation for the first time to American companies. But these imperialistic maneuvers in the Middle East simply encouraged incipient nationalism and resulted in the formation of the Organization of Petroleum Exporting Countries (OPEC).

Rising nationalist tendencies were an important factor in persuading the big international oil companies to diversify their operations, leading them into Southeast Asia, Alaska, and northern Canada. They also turned back into the North American continent, stepping up a campaign for more drilling on the Outer Continental Shelf and taking positions in other fuels, including coal and uranium.

Throughout the 1960s the energy industries underwent a quiet but important transformation. As in the past, it was carried out with the cooperation and assistance of the federal government.

The energy crisis itself was well advertised in advance by the oil companies, and as subsequent events made clear, was largely a contrived event. Viewed in a wider context, it was part of an overall reorganization, resulting in higher prices for oil products, sidetracking the introduction of alternative energy sources, opening up more of the coastal waters for drilling, and launching an important national debate on the industry's plans for a synthetic fuels business.

While most of the legislative "reforms" sounded like company public relations programs, there were signs of more serious resistance at local levels. In Georgia, Vermont, and Massachusetts, along the northern prairies, in San Francisco, ordinary people whose way of life is threatened by the reorganization of the energy industry are fighting back. In doing so, they have formed what may become the beginnings of an important new political coalition, bringing together environmentalists, with their historic concern for conservation, with consumer and welfare rights organizations, which have borne the brunt of the crisis in the form of higher electric and gas rates.

New Energy is a guide to the continuing crisis. It seeks to explain the immediate situation in terms of historic origins and to show why change, through either administrative or Congressional action, will be difficult to achieve. It goes on to describe efforts by local groups to regain control over energy and finally lays out a tentative plan for a decentralized energy system.

In writing this book, it is our hope to provide a new focus for the growing debate on natural resources, a debate we believe will dominate national politics for the remainder of this century.

PART ONE

The Roots of the Crisis
A Brief History of the Energy Industry

1. The Standard Trust

At the height of the energy crisis of 1973 grave doubts were expressed within and without government that the shortages were real. Even as citizens lined up for gasoline, Congressional investigators discovered the major oil companies had apparently cut back the operations of refineries, as if to create the shortages the firms insisted they sought to avoid. For a time the Arab boycott seemed to explain the shortage, but then it was revealed that the United States was importing as much oil in the winter of 1973 as it had the previous year when there was no boycott.

Senator Frank Church distributed a memorandum discovered in the files of the Standard Oil Company of California. Written in the late 1960s, it warned other oil companies of an oversupply and urged them to organize against market disruption. Testimony by officials from Aramco suggested that big company actually welcomed the Saudi Arabian boycott.

As Senator Church's agents delved into the history of international oil, they traced the development of an international petroleum cartel and followed the cartel's operations from the end of the Second World War, during which time the companies and the State Department had worked to organize Middle Eastern oil for U.S. interests. All of this had been carried forward in the greatest secrecy, without knowledge of Congress or the public.

Gradually it became apparent that there was no simple explanation for the energy crisis, that any serious interpretation would lead back into the 1920s, and in the end to John D. Rockefeller, who had created the modern industry by founding the great Standard Oil Company.

The Roots of the Crisis

In 1858 John D. Rockefeller pooled his savings with those of M. B. Clarke, an Englishman, to begin a produce commission business on the Cleveland docks. The business flourished, partly due to the Civil War. In 1862 Clarke and Rockefeller put $4,000 into a refinery run by Samuel Andrews, another Englishman. That enterprise burgeoned and soon was worth $100,000. Rockefeller then sold out his interest in the produce commission business, and in 1865 concentrated his efforts on the firm of Rockefeller and Andrews. Soon they opened another refinery and an office for selling oil in New York. Rockefeller was head of all the firms. Then in June 1870 Rockefeller merged the different companies into the Standard Oil Company with a capitalization of $1 million. The principals in the company were John D. Rockefeller, Henry M. Flagler, Samuel Andrews, Stephen V. Harkness, and William Rockefeller.

Rockefeller had a way with money:

> Among the early experiences that were helpful to me that I recollect with pleasure was one in working a few days for a neighbour in digging potatoes—a very enterprising, thrifty farmer, who could dig a great many potatoes. I was a boy of perhaps 13 or 14 years of age, and it kept me very busy from morning until night. It was a ten-hour day. And as I was saving these little sums I soon learned that I could get as much interest for fifty dollars loaned at seven percent—the legal rate in the state of New York at that time for a year—as I could earn by digging potatoes for 100 days. The impression was gaining ground with me that it was a good thing to let the money be my slave and not make myself a slave to money.

This principle then informed the Standard Oil Company.

Rockefeller built up the refinery business by obtaining secret rebates on oil shipped by the railroads. He ploughed the profits back into the business, in the process building up a large-scale cash reserve, which could enable the enterprise to survive in hard times. The oil business had peculiar risks.

There was a general fear that the oil would be exhausted, leaving the refinery owners stranded with useless equipment, and the fear that the reckless competition among the producers might make the business unstable and unfeasible. Rockefeller foresaw all this; and by putting back the money into his operation, he was able to survive hard times, to buy out other refineries, and to extend pipeline systems.

He accomplished all this with cunning and resourcefulness, promising the railroads he could provide them with a steady shipment of oil for a rebate. In fact, he did not have the oil. But with rebate in hand, he turned around and persuaded or browbeat other refiners into selling to him.

There was no beating Rockefeller. Confounded, tricked, and at the mercy of Standard Oil, independent producers formed a limited partnership called the Tidewater Oil Company for the purpose of bypassing the railroads by running a pipeline from the oil fields of Pennsylvania to Baltimore, where it could be moved by railroads independent of Standard. It soon became apparent that this pipeline would revolutionize the business. Oil could be transported by pipeline at 16 2/3 cents per barrel, as compared to 85 cents paid by Standard at its rebated rate.

Because of his policy of building up cash, Rockefeller was never in the position of placing his company in debt to banks as the railroads had done.

But even before the seaboard pipeline could be completed, Rockefeller was on the move, laying pipes throughout the oil fields connecting the oil fields to Bayonne and Philadelphia refineries.

The Tidewater pipe and rail system was meant to supply independent refiners. But Rockefeller, through intermediaries, approached those independents and bought them all out save one, which was condemned as a public nuisance and made to move.

Still the independents who formed the Tidewater struggled on. Securing a $2 million loan from the First National Bank, Tidewater began to build its own refineries. At first Standard sought to spread word in Europe, where the bonds were to be sold, that Tidewater was insolvent. When that failed, Standard quietly bought into the company and sought to disrupt stockholders' meetings.

Then Rockefeller changed tactics. In the beginning he had come as a cruel destroyer of the independent oil business. Now he meekly sought Tidewater's cooperation so that Standard and Tidewater could work together as partners. Thereupon Tidewater agreed to a fixed percentage of the business and Rockefeller made peace. The Tidewater originators kept 11 1/2 percent of the business, all the rest going to Standard.

By 1879 Standard Oil was the only buyer, only shipper, and only refiner of all but 10 percent of the oil in the country. In 1870 the cash resources of the company were $1 million; by 1875, $13 million; by 1881, $45 million. During that time the company not only bought its competitors but paid out dividends of $11 million.

Standard in effect had become its own bank; and beginning in 1882 with the creation of the Standard Oil trust at 26 Broadway, the company became the bank for the entire oil industry.

The men who ran Standard sought outlets in other investments. Through William Rockefeller, the brother of John D., the Standard Oil men became involved with the National City Bank. It occurred this way: In the early 1880s William Rockefeller was a director of the Chicago, Milwaukee, and St. Paul Railroad. In that capacity he met and became friends with James Stillman, a young New York cotton merchant. Later Stillman brought Rockefeller on as director of the Hanover Bank. Then, in 1891, Stillman became president of the City Bank, and from that time on the Rockefeller business gravitated toward this bank.

As a result, the bank's deposits grew from $12 million in 1891 to $31 million in 1893 and kept on growing until it was the largest bank in New York. The Rockefellers were not interested in the usual banking return of 5 or 6 percent on their money; they looked for more substantial returns. They quietly agreed to back Stillman, who in turn arranged to supply the funds for Edward Harriman's acquisition of the Union Pacific Railroad. This was a major investment, requiring $45 million in outright cash to pay off the government debt, and the City Bank was thought to be the only institution which could handle the transaction. This arrangement, then, brought the Rockefellers into dealings with Harriman, and the investment house of Kuhn, Loeb.

Both Stillman and Rockefeller became directors of the reorganized line. While at first the Rockefellers did not commit large amounts of money to the railroad, they, together with Harriman and Kuhn, Loeb, held one-third of the entire stock. In the following decade the Union Pacific became a veritable bank, piling up $1.5 billion of its own capital.

Meanwhile, the profits of Standard Oil grew. William Rockefeller expanded his interests in the Milwaukee railroad. John D. Rockefeller joined with the estate of Jay Gould in making large investments in the New Haven, Hartford and other eastern railroads. The City Bank and H. H. Rogers of Standard Oil formed the Amalgamated Copper Company, which bought out the old Anaconda Copper Company. The Rockefellers bought the Consolidated Gas Company, then merged it with competitors who were associated with the City Bank, and finally combined those companies with Edison Illuminating Company, thereby bringing all the lighting companies in New York under their control.

By this time the City Bank was the largest financial institution in New York, with deposits over $100 million.

These, then, were the origins of the Standard trust and the companies that depended on it.

2. The Modern Industry

In 1911 the courts in an antitrust action broke up the Standard trust into 33 different companies. These companies survived the dissolution and became important enterprises in their own right. In 1925, for example, the successor companies to Standard Oil controlled 47.4 percent of proven oil acreage in the United States.

Rockefeller had built his oil empire through control of refining and marketing. In order to operate in such circumstances, those companies that did compete with the original trust had integrated themselves, that is, established control over production, transportation, refining, and marketing. Among them were the Texas Company, Pure Oil, Union, and Gulf.

After dissolution the Standard offshoot firms moved in, in turn, to integrate themselves. Thus, Standard Oil of New Jersey merged with Humble Oil and Refining Company. That merger gave Standard control over valuable crude oil supplies and refined products for eastern markets. Standard Oil of California merged with Pacific Oil, the largest crude producer in the United States at the time, and so on.

As the petroleum industry matured, the major companies established understandings with both state and federal governments, so that government became a major instrument in ordering the petroleum business. Essentially the government reacted to the interests of the major companies, institutionalizing through state apparatus policies the companies found difficult to implement themselves. This was especially true in organizing a method for regulating the flow of crude oil that threatened to ruin the industry. Slowly government became the industry's agent in determining how much oil was to be produced out of the ground, first on the state level, later at the federal level. This was always done and it is today, in a spirit of coordination, to facilitate the operations of the major companies.

The major oil companies argued for "conservation" of oil as grounds for state and even federal control over production. More often than not this was a deliberate confusion in terms. Strictly speaking, conservation meant avoidance of waste in the recovery or use of oil. But what the companies meant by conservation was limiting production so as to control markets, avoid oversupply, and hence maintain price levels.

Reserves

By the late 1930s, the major petroleum corporations had obtained extensive reserves. The Temporary National Economic Committee (TNEC) estimated that the major companies held 70 percent of the then proven crude oil reserves as reported by the American Petroleum Institute. The most significant holdings were by Standard Oil (New Jersey), The Texas Corporation, Gulf Oil, and Socony-Vacuum, which together had 32 percent of the total reserves.

At first the major companies sought to limit the flow of oil through production policies. There were numerous indications then that they held their reserves off the market. At the committee hearings, E. DeGolyer, a leading oil geologist and witness for the American Petroleum Institute said, "Whether by force of circumstance or design, the big companies are able to market their reserves less rapidly than are the small companies and individuals." He pointed out that while Jersey Standard had 2.5 billion barrels of reserves, "they are being produced at approximately 40 percent of the rate averaged for the rest of the nation's production."

It was established that while independent oil operators discovered about twice as much oil as the majors, the majors held approximately 70 percent of the proven crude oil reserves. This situation arose because the majors prevented the independents from drilling minority leaseholds.

According to the TNEC, the system worked like this:

> An individual owns a small lease which shows on the major company's map as being a probably productive area. He will then be approached by a representative of the major company who will probably offer a higher price than they have been paying for leases before that time. If the independent owner will not sell at these terms, an effort is made to trade him a certain number of royalty interests for his lease. If necessary, he will be offered a royalty interest in a better position on the structure than his lease. Until a few years ago, when enforced unitization began to be used, it was customary for the majors to pay finally whatever price the relatively small lease appeared to be worth, based upon the value of acreage which by that time might have been developed.

Their primary aim is to lease land as rapidly as possible after it is discovered and to make every effort to control its production so that the best possible price can be obtained. As long as a small company has a lease on the structure, it is difficult to hold these reserves.

Teapot Dome

But these efforts were unsuccessful, and the big companies turned to government to help them stem the flow of oil. Meanwhile liberals and conservationists were out to regulate the industry.

During the Wilson administration Josephus Daniels, Secretary of the Navy, wanted the government to own and operate an oil industry to fuel the Navy. But his idea got nowhere; and it was partly in reaction to Daniels that soon after Albert Fall became Secretary of the Interior in Harding's Cabinet, he persuaded the President to transfer control of the naval oil reserves from the Navy Department to Interior, where he could lease them out to private industry. In June 1921 Fall leased the two most important reserves:

Elk Hills in California was awarded to his friend, E. L. Doheny the oil tycoon; Teapot Dome in Wyoming went to Harry Sinclair.

Supporters of Daniels, including Senator Robert La Follette and Gifford Pinchot, campaigned against Fall's program. The press made a fuss, and La Follette was persuaded to launch an investigation. It was then revealed that Doheny had kept prominent politicians in his pay, among them Franklin K. Lane, Wilson's Secretary of the Interior, and William McAdoo, Wilson's son-in-law and the former Secretary of the Treasury. It was also disclosed that Fall received $404,000 in loans from Doheny and Sinclair. Subsequently Fall was convicted of receiving a bribe and sentenced to a year in jail.

With the Teapot Dome scandal in the news, and the industry for its own reasons anxious to curb oil production, President Coolidge established in 1925 the Federal Oil Conservation Board, made up of the Secretaries of Commerce, the Interior, the Navy, and War. It was meant to cooperate with and assist the industry, which warmly welcomed its creation. This board was the forerunner of other federal agencies and established the basic relationship between the government and the industry. It conducted research into alternative fuels, laid plans to avoid shortages, and made various investigations. In 1926 Congress sought to assist the oil companies by adopting a 27.5 percentage depletion allowance on oil.

This early agency and the liaisons it fostered reflected the thinking of Herbert Hoover, Secretary of Commerce under Coolidge, who believed that industry and government should form cooperative alliances for mutual advantage. But when the American Petroleum Institute sought to put Hoover's theories into practice by proposing a cooperative scheme among the major companies to withhold production (i.e., a cartel), Hoover, by then President, came down against the idea. The API then abandoned the plan for a federally sanctioned domestic cartel, and instead turned to the states for help.

Prorationing

The states created various prorationing laws, which in the name of conservation established production levels. These prorationing laws gained federal sanction under the Roosevelt administration. The National Recovery Administration law prohibited the production of oil in interstate and foreign commerce in excess of the amount permitted by the proration laws. (The oil code under the NRA also provided for the limitation of imports of crude oil, for restrictions on the withdrawal of crude oil from storage, for periodic estimates of consumer demand, and for the allocation of production among pools in states.)

Before the NRA was invalidated by the Supreme Court, Congress passed the Connally Act in February 1935 as a substitute for the oil section of the NRA. It specifically prohibited the movement in interstate commerce of "hot oil," that is, oil produced in excess of the state proration quotas.

The effect of the prorationing laws was to drive out the independents. The TNEC reports what happened when the big East Texas field was discovered in 1930. Between 1930 and 1938 over 155 independent refineries were built in this field, only one by a major company. By 1940, there were but 3 independent refineries left. The independents were eliminated with the institution of prorationing. Prior to prorationing they could obtain most of their oil from their own wells, but with prorationing (that limited the amount they could take from their own wells), they had to buy oil on the market.

The price the refiners paid for crude oil and received for the sale of gasoline was determined by the posted price of the majors. In order to rid themselves of independents, the major companies applied the so-called "price squeeze," by posting or setting the price of crude oil high while maintaining the price of gasoline at a low level.

Transportation

Transportation was crucial to the control of oil production, even more than in the days of Rockefeller. The TNEC

The Modern Industry   11

report showed 14 major oil companies controlling 89 percent of the pipelines carrying oil and oil products. By then virtually all oil moved by pipelines.

The major companies continued the rebate policies artfully elaborated by Rockefeller. They joined together in ownership of pipelines through subsidiaries. The subsidiaries paid the major companies rebates in the form of dividends. The rate of return on a pipeline investment was 26.7 percent. Even during the depression the earnings of pipelines subsidiaries never declined. The majors could take a loss in refining or marketing, thereby holding down threats from independents, because they controlled the pipelines and took their profits that way.

Partly as a reaction to the Standard trust, the Hepburn Act of 1906 made pipelines common carriers and required them to transport products of third parties. But that turned out to be little help to independents. Before the Hepburn Act, the Standard trust refused to ship products by independents. After the act's passage, the Standard companies required that independents ship at the rate of 100,000 barrels a day. By continuing to establish high rates for minimum shipments, these companies discouraged independents from using the pipelines.

But even where they could use a pipeline controlled by a group of majors, independent companies had to pay a high price. The TNEC provides this example:

> Standard Oil Co. (Indiana) owns the Stanolind Pipe Line Co., which extends from fields in Oklahoma and Texas to the parent company's huge mass-production refinery at Whiting, Indiana (near Chicago), a distance of over 500 miles. During 1938 the Stanolind Pipe Line Co. transported 43,485,625,000 barrel miles of crude oil at a cost of $11,050,478, which included all operating expenses, State and Federal taxes, and fixed and contingent expenses. This is an average cost of only 0.032 cents per barrel mile. An examination of the company's tariff filed with the Interstate Commerce Commission discloses that the rate from Oklahoma to Whiting, Indiana, was 34.5 cents per barrel, or 0.069 cents per barrel mile based on

500 miles. This shows unquestionably that the cost is less than half the tariff rate which must be paid by independents if they do ship over the pipeline.

Mr. Louis J. Walsh, an independent refiner of Texas, testified before the Temporary National Economic Committee that it costs 17 1/2 cents per barrel to get oil from the East Texas field to the Gulf Coast by major pipe lines, but the cost to the majors is only 5 cents per barrel.

Refining

The major oil companies achieved much of their market dominance through control of refining operations. The simple process of a refinery is to place crude oil in a tank and then apply heat. A vapor is given off, which is passed through condensers. Different products pass through different holes. Gasoline, which is the lightest, is given off first. As refining developed, the processes became more complex. The cracking process, which enables the refiner to break down crude oil under heat and pressure, made it possible to produce more gasoline from crude oil than before. In 1920 the recovery of gasoline was 26.06 percent. In 1939 it was 44.9 percent.

The oil companies developed refining centers. For instance, in 1940 there were 547 refineries in 34 states. But 10 states had 90 percent of the capacity. Texas and California together had 50 percent of the total. The major companies owned most of the big refineries, which produced many different products. By 1930 they had bought out all the independent refiners on the East Coast.

The major companies had 75.6 percent of crude oil refining capacity in 1938. They all owned cracking plants, which represented 85.2 percent of the total. The independents who owned cracking plants had to pay royalties on patents controlled by the majors. Six major companies owned 45.2 percent of the crude oil capacity and 53.5 percent of the cracking capacity.

Gasoline Marketing

Patent control was important in the developing gasoline business. The major oil companies basically held patents through jointly held patent subsidiaries. The majors seldom opposed one another in patent litigation, and generally they purchased patent rights from one another. While it was rare for major companies to sue one another over patents, the big companies frequently used their patent control to harass independents.

We have already seen how the East Texas field was brought under control by the majors, and the independents squeezed out. There were other schemes by which the major companies policed the gasoline markets. One practice was to buy at slightly higher than market prices so-called "distress" gasoline, or overproduction, and store it, thereby removing it from the market. In 1929 the large firms had in place the Pacific Coast cartel, a collective agreement among the major companies to maintain gasoline prices and to buy up excess production from independents if they agreed to maintain prices mutually agreed upon.

In the Midwest a buying program coordinated by Socony-Vacuum had the majors buy up distress gasoline and then control the price through control of the tank car rates and other means. The ultimate aim was to raise the price to consumers.

Most of the major companies exchanged gasoline. The principle is that a major supplies other majors gasoline for their marketing outlets that are near its own refinery in return for gasoline needed at its own marketing outlets located at distant areas. In 1937 over 96 percent of the gasoline received by major companies was from other majors on an exchange basis.

The most important growing market for petroleum was gasoline; and it was effectively manipulated through the Ethyl Gasoline Corporation. Ethyl manufactured a product called tetraethyl lead, which was mixed with gasoline to raise its antiknock qualities. All majors used the fluid. But independent jobbers and refiners could not obtain it unless they agreed to a pricing policy stipulated in the contract. That stipulation said premium gas must be sold at 2 cents more per gallon than regular gas, even though the real cost of the tetraethyl lead was only .037 cents per gallon. Ethyl Gasoline was owned 50 percent by Standard Oil (New Jersey) and 50 percent by General Motors. During this period GM in turn has owned 23 percent by the Du Pont interests.

Rockefeller Family Interests

The Rockefeller oil industry holdings continued to be significant. The TNEC report shows that as of 1938 the market value of the Rockefeller holdings in the 200 largest nonfinancial corporations was $397 million. That represented 1.5 percent of the total market value of all the stock. Most of the family holdings were devoted to oil.

Stockholdings that conveyed control were:
- Standard Oil (New Jersey). (Between members and foundations, Rockefeller controlled 13.5 percent of the common stock. If the holdings by Indiana Standard in Jersey Standard are taken into account, the family holdings amounted to 20.2 percent.)
- Socony Vacuum. (Family holdings were 16.3 percent.)
- Standard Oil (Indiana). (Family and foundation holdings were 11.4 percent.)
- Standard Oil (California). Family and foundation, 12.4 percent.)
- Ohio Oil Company. (Family holdings 18.6 percent.)

By the beginning of World War II, the Rockefeller interests had grown larger, not smaller, and their influence more pervasive. And rather than organize the petroleum business in any sort of public interest, the federal government under both Republican and Democratic administrations had institutionalized the practices of the major petroleum companies.

3. The International Petroleum Cartel

Perhaps the single most important factor in the creation of the modern international petroleum industry was the decision by the British to fuel the fleet with oil instead of coal.

In 1901 William Knox D'Arcy had obtained a 60-year concession to explore for and produce petroleum "throughout the whole extent of the Persian Empire," an area of 500,000 square miles. D'Arcy did not strike oil, and when his funds were exhausted, sought additional capital in London. Failing, he undertook negotiations to sell the concession to foreign sources.

At this point the Admiralty entered the picture. Since 1882 naval officers had argued that the fleet could increase its fighting capacity by changing to oil. In 1904 E. G. Pretyman, chairman of the Admiralty's oil committee, approached Burmah Oil Company and encouraged that company to join with D'Arcy, putting up capital through a syndicate. In the meantime the Admiralty asked D'Arcy to break off talks with the foreigners. A syndicate was duly formed called the Concessions Syndicate, Ltd., and in 1908 D'Arcy struck one of the largest oil fields in the world. In that year Concessions Syndicate became the Anglo-Persian Oil Company (later the Anglo-Iranian Oil Company), with Lord Sratchona of Burmah Oil as chairman and D'Arcy a director. The British government, which informally had caused the company to come into existence, now sent troops to protect its operations in Persia.

When Winston Churchill became First Lord of the Admiralty in 1911, he set out to expand the Navy, making it more dependent on oil. With no oil supply in the British

15

Empire, the Admiralty turned to Anglo-Persian, which by this time had oil wells, a pipeline, and a refinery in operation. In 1914 the government both made a long-term supply contract and agreed to buy a controlling interest in the company. In proposing these arrangements, Churchill argued that they would free the Navy from reliance on Standard Oil and Shell, the world's dominant oil companies, and help insure a stable, inexpensive supply in times of both peace and war. Of the Persian concession he said, "Over the whole of these enormous regions we obtain the power to regulate developments according to naval and national interests, and to conserve and safeguard the supply of existing wells pending further development."

While the British were negotiating the Anglo-Persian agreement, the government was also taking a hand in shaping the oil business in Mesopotamia (now Iraq). During the period 1904 through 1914, when the British government was considering the change from coal to oil, the German and British interests were competing for what then looked to be important oil fields in Mesopotamia. The first reports on oil in Mesopotamia were made in the 1890s by C. S. Gulbenkian, the American trader. They aroused the interest of the Turkish Sultan, who transferred the lands from the Ministry of Mines to the Liste Civile; in fact, a transfer to his own private account. But the British and German negotiations were futile. They haggled unsuccessfully with the Sultan until 1909, when he was swept out of power by the Young Turk Revolution. In an atmosphere more conducive to the British, the National Bank of Turkey was created to promote British interests. In 1911, through the mediations of Gulbenkian, there was formed the African and Eastern Concessions, Ltd., which a year later became the Turkish Petroleum Company, and in 1929 the Iraq Petroleum Company, Ltd. (IPC).

As a result of the Admiralty's new oil policy, interests in Iran and Iraq were brought together. In 1914, through efforts of the British Foreign Office, Anglo-Persian and the Iraq

Petroleum Company agreed to a reorganization of Iraq Petroleum in which the D'Arcy group received a 50 percent interest. The other stockholders were Deutsche Bank and Anglo Saxon, each with 25 percent. The two latter groups agreed to contribute an amount equal to five percent for Gulbenkian.

At this point both the British and German governments pressed the Turkish government to lease lands for oil exploration, and by the outbreak of the First World War an agreement in principle had been reached.

Following the war, the French claimed the 25 percent previously held by German interests as war booty. And, in 1920 the San Remo agreement awarded the German share to the French, who in turn agreed to construct a pipeline to the Mediterranean. At this point the American companies entered the picture.

Since the early part of the century, the U.S. companies, led by Standard Oil Company and subsequently its successor firms, dominated the world oil markets. In 1921 the American oil industry produced about 65 percent of the world's oil supplies and purchased about 17 percent of the remainder, mostly from Mexico.

Standard's ability to maintain a dominant position without controlling oil fields was due to the continual discovery of new fields, which kept domestic production at high levels so that oil could be exported. This system worked all during the first 35 years of Standard's history.

But in the 1920s the successors to the Standard trust changed their attitude toward reserves in general and especially toward foreign reserves. This was due to a number of factors. In part, the companies were well aware of the big foreign finds. These prospects were now more attractive than they once might have been because cheap oil reserves in the United States apparently had run out; and securing additional reserves for expanding domestic markets meant purchasing oil lands from private holders at prices higher than those charged by the government in public domain territories. Thus, obtaining oil reserves necessitated tying up capital over many years.

More important, there was a growing fear of an oil shortage in the United States. An industry source describes this view as follows:

> Fear of an oil shortage in the United States was uppermost as a factor in international relations after World War I. It was a holdover fear from a narrow escape from scarcity in 1917-18 in the midst of war. It was fanned by what might have been an actual—although probably short lived—shortage had the war, with its tremendous demands on American supplies, been prolonged into 1919 and 1920. That it grew into a case of national jitters is not wholly surprising in view of the fact that the military importance of oil in modern war had been demonstrated. Oil supply took on a vital national defense complexion.

In 1920 David White of the U.S. Geological Survey said:

> On the whole, therefore, we must expect that, unless our consumption is checked, we shall by 1925 be dependent on foreign oil fields to the extent of 150,000,000 barrels and possibly as much as 200,000,000 barrels of crude each year, except insofar as the situation may at that time, perhaps, be helped to a slight extent by shale oil. Add to this probability that within 5 years—perhaps 3 years only—our domestic production will begin to fall off with increasing rapidity, due to the exhaustion of our reserves.

Another factor was the fear of foreign monopoly. A Senate investigation in 1920 indicated that American companies were systematically being shut out of foreign oil fields. At the same time Royal Dutch Shell, partially owned by British interests, was expanding rapidly within the United States, exploiting American resources while conserving oil elsewhere. Resentment was rife. In 1919 E. MacKay Edgar wrote in *Sperling's Journal*

> America is running through her stores of domestic oil and is forced to look abroad for future reserves.... The British position is impregnable. All the known oil fields, all the likely or probable fields outside the United States itself, are in British hands or under British management or control, or financed by British capital. . . . To the tune of many million pounds a year, America before very long will have to purchase from British companies . . . a progressively increasing proportion of the oil she cannot do without and is no longer able to furnish from her own stores.

Led by the Standard Oil Company of New Jersey, the American companies waged a vigorous campaign to participate in Middle Eastern oil. This was carried forward both through the State Department, where it took the form of diplomatic disputes for oil concessions in mandated territories, notably Mesopotamia, where the British were seeking to exclude American companies, and in secret initiatives by the company which sought to purchase Middle Eastern concessions.

In June 1922, the president of Anglo-Persian telegraphed the president of Jersey Standard asking him to meet with officials of the Turkish Petroleum Company. These negotiations then set the stage for the formation of an international cartel. After notifying the State Department, Jersey Standard commenced negotiations in behalf of seven major American oil companies. These negotiations began in July 1922 and lasted six years, until a final agreement was reached in 1928.

At the beginning of the discussions, the American companies placed great emphasis on what was called the "open door policy," and insisted that acceptance of this policy was basic to their entering into any agreement. The open door policy was interpreted to allow any company to obtain oil concessions in areas mandated under the League of Nations, including Mesopotamia. Theoretically the idea was to promote active competition for oil rights and to prevent monopoly of oil by any one company or group.

In 1928 the American companies were formally admitted to the Turkish Petroleum Company which by that time had changed its name to Iraq Petroleum Company. And thus, the major international companies were joined together for the first time. They included Anglo-Persian, Royal Dutch-Shell, Standard Oil of New Jersey, and Socony-Vacuum (now Mobil). The two American companies operated as a single unit through the near East Development Corporation. This then was the cartel.

During the period of negotiations and between the time the Americans had argued for an open door until the point they entered the IPC, the world had gone from a period of shortage to one of oversupply. Instead of competing for more oil, the international companies sought ways of limiting production and allocating world markets.

Within the United States the American Petroleum Institute was advancing its plan for a cooperative industry agreement that would have curtailed domestic production. This plan was denied on antitrust grounds, but it was later in large part implemented through ad hoc adoption of state prorationing laws. These laws gained federal sanction during the New Deal.

Even before the Americans entered, IPC undertook to tightly control oil in Iraq. It made a concession agreement with Iraq which made it impossible for any nonmember of the company to obtain concessions in areas that were to be opened to competitive bidding. The bidding was changed from public to sealed bidding with Iraq Petroleum opening bids and making awards. Although originally forbidden to bid on competitive areas, IPC in the concession agreement was allowed to outbid any outsider. All proceeds from the bidding were to go to the company.

In 1928 the different groups in IPC signed the so-called Red Line agreement, named for a red line that circled oil territories within IPC's purview, and agreed not to be interested in the production or purchase of oil within the area of the old Turkish empire (Turkey, Iraq, Saudi Arabia, and adjoining sheikdoms, except Kuwait, Israel, and Trans-

Jordan), other than through IPC. A concession agreement of 1931 further tightened the lines of the arrangement, shutting the door on outsiders within a large area of Iraq.

The British, American, and Dutch interests worked to retard the development of oil in Iraq. They used various tactics, including requests for extensions of time in selecting drilling sites, delay in constructing a pipeline, reductions in drilling, deliberate drilling of shallow holes with no intention of finding oil. Throughout the 1930s these interests sought to effect cutbacks in Iraq production, opposing enlargement of the pipeline on the grounds that increased production would upset world markets.

As time went on, the Red Line Agreement, whose major advocates were the French and Gulbenkian, became an albatross around the necks of the American participants in the cartel. The Red Line Agreement did not affect outsiders. Thus, the IPC watched as the British Development Corporation, an independent company, obtained a concession in the Mosul section of Iraq and Standard of California won concessions in Bahrein and Saudi Arabia. Iraq Petroleum interests dealt with the British Development venture by secretly buying up the company's shares and thus gaining control. They then sought accommodation with Standard Oil of California, in effect seeking a partnership that would include Standard within the cartel. But this proved difficult because neither the French nor Gulbenkian were willing to waive their rights which entitled them to a pro rata share of production within the area. When World War II broke out, the groups in the IPC had reached an agency agreement for purchasing among themselves Bahrein production, and thus containing the world market. But the war interrupted the plan.

After the war the American companies boldly shelved the Red Line Agreement, declaring it to be invalid because some of the former partners had been enemies and because it was a violation of U.S. antitrust laws.

At that time Standard Oil of New Jersey and Socony were anxious to buy into Aramco, a U.S. concession in Saudi Arabia.

When the American companies broke the Red Line Agreement, the French filed suit to enforce the agreement in British courts. But before the suit could come to trial in 1948, a new agreement was reached. Under the new agreement groups were free to act individually or in ventures with others within the Red Line areas. The different groups were free to obtain crude at arbitrarily low prices, and the British, Dutch and Americans maintained their control over Iraq Petroleum's management and policies.

The French, in a confidential document, summed up the purpose and function of the IPC: "The incorporation of the IPC and the execution of the Red Line agreement marked the beginning of a long-term plan for the world control and distribution of oil in the Near East."

4. World War II

During the Second World War, the oil industry and the federal government developed closer ties. Between 1924 and 1939 all efforts had gone toward restricting the production of oil in order to control supply. But during the war, industry and government worked hard to maximize production.

In the early days of the war, German submarines cut the Allied tanker traffic in the Atlantic and forced curtailment of crude oil shipments from California and Texas to the East Coast. Since the Atlantic Coast received 90 percent of its oil by sea, it faced serious shortages in 1940 and 1941.

Roosevelt asked Congress for authority to allocate oil production and distribution, but it refused to do so. The President then gave Harold Ickes, Secretary of the Interior, executive authority to arrange volunteer rationing systems and help facilitate the oil supply. Ickes worked at finding needed oil and arranging for its supply through scarce railroad tankage.

Because transportation was short, Roosevelt supported a plan to build a private pipeline system that would carry needed oil from the Gulf Coast to New York. But the Georgia legislature, at the behest of railroad interests, blocked the plan, refusing to give rights of way. In the end, the pipelines were financed by the federal government. The Big Inch ran from Texas to New York; the Little Inch from Illinois to the Atlantic.

As Petroleum Administrator for War, Ickes became closely involved with the industry. He persuaded the Justice Department to abandon antitrust enforcement, allowed the companies to enter into voluntary pooling agreements, and became a strong defender of preferential oil tax policies. And he sought to stimulate the industry by providing financial incentives to drillers, and long-term contracts.

In effect, the federal government widened its protective shield around the industry, spending billions of dollars to build a national transportation network for petroleum, developing new products, and encouraging the industry to work closely together by abandoning antitrust enforcement. At the same time, Ickes' efforts to make the federal government a partner and dominant force in the industry failed. The industry not only maintained control over operations affecting its welfare but used federal funds to accomplish its expansion.

This was clearly demonstrated in the Middle East, where Roosevelt moved to protect and expand the operations of Standard Oil of California and Texaco, the two American companies that owned Aramco, the most important U.S. concession. In 1941 King Ibn Saud asked Aramco for a $6 million advance against royalties. The company, then $34 million in debt on its investments there, instead gave him $3 million. Its officials then persuaded Roosevelt to advance Saudi Arabia money through the lend-lease program by way of the British government. As a result, Aramco payments to Saudi Arabia declined from $2.9 million in 1940 to $79,651 in 1943. Meanwhile, British loans to Saudi Arabia, backed by lend-lease, rose from $403,000 in 1940 to $16.6 million in 1943.

Although Aramco officials at first had begged for U.S. aid to Saudi Arabia, by 1943 they feared that the British had used the lend-lease funds to influence the King and would attempt to remove Aramco from the concession. Once more, then, the Aramco officials went to Washington and persuaded Ickes and others in the Roosevelt administration to give lend-lease aid directly to Saudi Arabia on the grounds that the country was vital to U.S. interests.

Ickes had repeatedly argued for a centralization of the oil industry, and in 1943 he proposed to Roosevelt the formation of the Petroleum Reserves Corporation, which could acquire and participate in the exploitation of foreign oil reserves. The first job for the corporation would be to acquire participation in the Saudi Arabia concession, thereby forever eliminating the British. Roosevelt set up the corporation and made Ickes chairman. But when Ickes told Aramco representatives of his plan, they were furious and promptly set about turning the industry against him. Negotiations took place, with the government first asking for a 70 percent interest in the business. This demand was then dropped to 51 percent, then to 33 percent, and finally, because of opposition, abandoned altogether.

Ickes made one more attempt to establish the U.S. government in the Middle Eastern oil business. He seized on a Navy plan that the government should build a pipeline from the Persian Gulf to the Mediterranean. The government would build the pipeline and in exchange would get oil at 75 percent of the market price. The Aramco officials supported this plan, for it provided new markets for their oil at no cost; and because the government was involved, the pipeline was sure to be built speedily. But the thought of government intervention was too much for the industry, and the scheme was beaten.

Instead Aramco announced it would build the pipeline on its own. At the end of the war, with great demand for materials to rebuild national economies, the United States granted Aramco 20,000 tons of steel to construct the pipeline; and through its previous allocations, it allowed the company to proceed not only with the trans-Arabian pipeline but also with a new refinery and other pipeline projects.

While these maneuvers were going forward, Aramco was working out new markets for its oil. Until 1941 Caltex was able to find markets east of Suez for only 12,000 to 15,000 barrels of oil a day, only one-seventh the amount that Aramco could produce. During the war, production was increased to 58,386 barrels per day by 1945. But a large amount of this oil was refined and sold to Allied governments for the war, and it was thus only a temporary outlet. By the end of the war, Aramco was left with crude oil and refineries but no market. Aramco's need for markets was made acute by discoveries of new fields in 1945 and 1947. It was at this point, and for the above reasons, that Aramco went ahead on its own to build a pipeline to the Mediterranean, thereby opening European markets.

As a result, the other established oil companies sought to open up markets for Aramco, while at the same time stabilizing the oil business and avoiding oversupply. This involved several different steps. First, the Texas Company sold its European marketing operations to Caltex, thus making a market west of Suez available to Aramco. Then Standard of California and Texas allowed Standard Oil of New Jersey and Socony Mobil together to buy 40 percent interest in both Aramco and the trans-Arabian pipeline. Third, Jersey Standard and Socony contracted to buy oil from Aramco.

The same sort of division took place in Kuwait, where Anglo-Iranian and Gulf shared the concessions of the Kuwait Oil Company on a 50–50 basis. The agreement stated that neither company would use the Kuwait oil to upset the other's marketing operations and that Anglo-Iranian specifically reserved the right to substitute oil it produced in Persia or Iraq to supply Gulf's additional requirements. Although oil was discovered in Kuwait in 1938, lack of transportation and facilities postponed development of the oil until after the Second World War.

In effect, the cartel was reinforced and extended through a series of long-term supply contracts in which companies with surplus crude production supplied oil at cost to crude-short companies in return for an even split of the profits. Competition between the companies at the producing and marketing ends was thus precluded. Gulf agreed to supply Shell over 10 years with supplies ranging from 1 to 10 million tons of crude per year, with all costs and profits pooled and divided equally. Jersey Standard and Mobil agreed to purchase 134 million tons over 20 years from Anglo-Iranian in Iran and/or Kuwait.

The extension of the cartel, of course, was well known in Washington, where it was supported by policymakers. A February 1947 memorandum from the Assistant Chief of the

State Department's Petroleum Division makes clear the effects of these policies:

> It is likely that with the dropping off of military demand there was no close relationship between the war-developed productive capacity of these companies and their peace-time market outlets. There can be no question but that this was true if the comparison were between their relatively small peacetime market position and their existing and potential production based upon a reasonable program of development. Thus, with these companies holding vast reserves pressing for market outlets, the prospects were favorable not only for increased competition for markets between them and the principal established marketers—Shell, Jersey, and Socony—but also for the development of a competitive market for crude oil in the Middle East upon which other companies might rely for supplies in entering the international oil trade.
>
> Under the three large Middle East oil deals, a major change in the conditions favorable to competition would seem to be in prospect. Instead of undertaking to develop their own market outlets for their present and prospective production, Gulf, Anglo-Iranian, and Arabian-American have aligned themselves through long-term crude oil contracts and partnership arrangements with the three large marketing companies, thereby at least contributing toward if not in fact precluding, the development of any bona fide competition for markets between the two groups of companies....
>
> ... the question naturally arises as to whether some other feasible pattern of ownership of Middle East oil resources would not be in the public interest and more consistent with present United States foreign economic policy. It is believed that from these standpoints, the case is strong for doing what may be feasible toward making ownership of Middle East oil possible for a larger number of companies. It is believed that from the same standpoints the case is strong for discouraging, wherever possible, the further development of joint operations and joint interests between and among the large international oil companies.

Oil Under Truman

After the war J. A. Krug, the new Secretary of the Interior, hoped to incorporate wartime experience into postwar policies. The Petroleum Administration for War was dissolved. Krug sought to establish coordinating links between the industry and government. He formed the Oil and Gas Division in the Interior Department, which was meant to function as a communications link between state regulatory agencies and the federal government, as well as a clearinghouse for oil and gas statistical and technical data. Krug wanted to establish an industry advisory board; and at his request Attorney General Tom Clark formally noted that an industry advisory board would not be in violation of the antitrust laws and that consultation in such a group would not be deemed a violation of the law. This, then, led to the creation of the National Petroleum Council, which includes chief executives of most of the major oil companies, refiners, producers, and other segments of the industry.

Tidelands Oil

While Truman did not tinker with the basic shape of the industry, he did take a strong position against control by the states of the tidelands oil. Ickes had wanted the tidelands under federal control. In 1937 the Senate passed a bill to that effect, but the House failed to act. The war prevented Ickes from further pursuing this objective. But after the war he approached Truman on the issue, and on September 28, 1945, Truman issued a proclamation announcing that the U.S. government regarded natural resources in the subsoil of the seabed of the Continental Shelf as federal property. He vetoed a Congressional resolution that would have awarded the tidelands to the states, and he instructed the Attorney General, Tom Clark, to file suit in the Supreme Court against California, which sought ownership. The Court denied the state control. Between 1948 and 1953 there were numerous bills in Congress. In 1952 both Senate and House approved a compromise bill, which Truman sent back with a conciliatory message. Tidelands oil became a major issue in the 1952 election. After the election Congress passed two bills: The Submerged Lands Act of 1953 gave to the states oil found under territory three miles out. Oil found further out on the Continental Shelf was under federal jurisdiction, and under the Outer Continental Shelf Act the government could lease that oil.

5. Oil and the Cold War

For a brief moment towards the end of the Truman administration, the international petroleum cartel was brought under frontal attack by the Antitrust Division of the Justice Department, which sought to break it up. But the State Department turned aside the attack, and in close liaison with the major oil companies staged a series of diplomatic initiatives that had the effect of maintaining the cartel in place and even extending its scope, making it in effect an instrument in Cold War policy.

The Middle East was the scene of the first confrontation between the Soviet Union and the United States following the Second World War, and oil was inextricably tied in with the international politics of the period.

George McGhee, the Assistant Secretary of State for Near Eastern Affairs in 1949 has testified:

> At that time the Middle East was perhaps the most critical area in the world in the contest between ourselves and the Soviets. The governments in the area were very unstable. We had no security pact covering this area. The Soviets had threatened Greece, Turkey and Iran. As a result of a very strong position taken by President Truman, we were able to dislodge the Soviets from northern Iran where they had demanded an oil concession. Although we had already bolstered Greece and Turkey through the Greek-Turkish Aid program, both were still in a precarious state.
>
> The Arab States were very hostile to us because of our involvement in the Israeli affairs. Saudi Arabia was more tol-

erant than the others, King Ibn Saud always seemed a little less affected by Israel in his relations with us than the other Arab States. There were, however, threats of strikes against us in Saudi Arabia. Always in the background there was the possibility of some nationalist leader, particularly in the countries where there were kings and sheiks, who might seize power as Nasser did later.

I think in retrospect this was always the greater danger in the Arab States than communism itself, which didn't find fertile ground among the Arabs. It was only later through Egypt that the Soviets obtained access to the Arab states.

At this time the principal threat to the Middle East lay in the possibility of nationalist leaders moving to upset regimes which were relatively inept and corrupt, and not attuned to the modern world.

There was also always in the background the reaction in the Arab states to what happened elsewhere. For example, had there been a communist seizure in Iran, we would have expected a similar threat in the Arab states.

The oil concessions themselves were of immense direct value to the United States. As McGhee, a former oil producer, put it: "The ownership of this oil concession was a valuable asset for our country. Here was the most prospective oil area in the world. Every expert who had ever looked at it said that this was the 'jackpot' of world oil. To have American companies owning the concession there was a great advantage for our country."

But the continued stability of the oil concessions was uncertain. In Venezuela the government had increased its oil revenues by raising the taxes on production. This policy apparently had an effect on Iran's Nationalist Premier, Mohammed Mossadegh, who demanded more revenue and participation in the affairs of the Anglo-Iranian Oil Company. Anglo-Iranian responded by offering a 50–50 profit-sharing plan, but Mossadegh refused. When the British company would make no further concession, Mossadegh nationalized it. In Saudi Arabia the Finance Minister threatened to shut down the Aramco concession if more money was not forthcoming.

Both the Aramco officials and the U.S. State Department, acting independently, concluded, as McGhee put it, that a "big move had to be made."

"We felt it exceedingly important from the standpoint of the stability of the regimes in the area and the security of the Middle East as a whole and the continued ownership of our oil concessions there and the ability to exploit them, that the Government of Saudi Arabia receive an increased oil income," McGhee said.

The U.S. government thereupon helped the companies negotiate a new policy under which the American oil companies agreed to pay the Arab governments higher royalties in return for lower taxes in the United States. Looked at another way, the oil companies became the vehicle for foreign aid to the Arab countries, in this instance without the knowledge, let alone the approval, of Congress.

The policy took shape in discussions between the State Department and the companies on a 50–50 profit-sharing formula, and between the Treasury Department and the companies on a tax program. The overall policy was then put together within the National Security Council.

As negotiations went forward, there were meetings at the State Department between government officials and company executives on the details of profit-sharing, and in 1950 Saudi Arabia and Aramco concluded a 50–50 arrangement.

It is misleading to describe this agreement as an equal sharing in profits. As George W. Stocking points out: "The agreement does not provide for the equal sharing of what in customary oil production accounting are regarded as profits. Under the original agreement Aramco paid the Saudi Arabian government annually a royalty of five shillings on each ton of oil it produced (except for what it used in its own operations). It appropriately treated this as a cost of operation in calculating such income taxes as it was required to pay under United States corporate income tax laws, and it continued to do so after the 1950 profit-sharing agreement. The 1950 agreement provided that Aramco pay the Saudi Arabian government one half of net profits after meeting its foreign tax obligations, and it permitted Aramco to treat all royalties, rentals, and other sums paid or payable to the government as credits toward the government's share of its net earnings. The payments to the government as a tax accordingly represented somewhat less than 50 percent of net earnings as customarily calculated."

The Treasury, meanwhile, decided to treat royalty payments to the Arab countries as taxes, at the urging of the State Department.

Before 1950 the oil companies had treated royalty payments as a business expense and had deducted them from gross income. A series of secret IRS rulings allowed the companies to treat royalties as taxes paid to a foreign government, and to deduct them dollar for dollar from their U.S. tax bills.

In effect, the oil companies agreed to pay the Arab governments higher royalties in return for lower taxes.

The taxes the four American oil companies that own Aramco paid to the United States fell from $50 million in 1950 to $6 million in 1951. At the same time, Aramco payments to Saudi Arabia went from $66 million to $110 million, a gain of $44 million.

"Wall Street lawyers were sent to the Middle East to help these countries rewrite their laws to bring them within the purview of the tax credit provisions of the United States Internal Revenue code," Senator Church said. His multinational corporation subcommittee investigated the origins of the tax arrangement.

In 1974 looking back on the arrangements he had helped to negotiate, McGhee said, "I believe the 50–50 was responsible for providing the stable basis for Aramco operations which enabled it to last 22 years, until 2 years ago. I don't know of any agreement that has provided a better basis for two parties to do business together. I consider this a great accomplishment for both the industry and the government."

Shortly after these arrangements were made and independent of them, the Justice Department pressed ahead with an attack on the cartel. On July 17, 1952, the Attorney General announced that a grand jury investigation of the international oil industry would begin. This investigation followed several months of debate within the administration and was based on a lengthy detailed staff report on the international petroleum cartel prepared that spring by the Federal Trade Commission staff.

But the State Department opposed this course. In a memorandum of April 16, 1952, Dean Acheson wrote, "In the view of the Department of State the institution of these proceedings will not help the achievement of the foreign policy aims of the United States in the Middle East and has the possibility of seriously impairing their attainment." He went on to point out:

> From the memorandum supplied by the Justice Department it is observed that one aspect of the alleged conspiracy involves control of the major oil-producing area in the world, particularly in the Middle East. This will inevitably be interpreted by the peoples of the region as a statement that, were it not for such conspiracy, they would be getting a higher return from their oil resources. This will, of course, strengthen the movement for renegotiation of the present concession agreements and may give encouragement to those groups urging nationalization. Since the issues are not only economic but also political, the net effect will probably be to cause a decrease in political stability in the region. It would also appear to be likely that the time required for an adjudication of the issue by the courts will be such that the effect will be the same regardless of what the courts ultimately decide.

Even so, subpoenas were issued in August and a grand jury was impaneled. Now, however, the cartel case became more intertwined with U.S. foreign policy. In March 1951 Iran had nationalized the assets of the Anglo-Iranian Oil Company, the majority of whose stock was owned by the British government. A bitter dispute between Britain and Iran ensued, with Britain threatening to sue any company that handled oil from the nationalized industry. (Actually, the decline in Iranian production was a boon to the other members of the cartel, which had a surfeit of oil. They simply increased their production elsewhere.)

For a time following nationalization, the American government apparently sought to negotiate a solution to the Iranian oil question by finding some way to organize Iranian oil production without the Anglo-Iranian Oil Company. Obviously, the takeover offered the possibility of American entry into Iranian oil production, hitherto controlled by the British. But as Acheson later wrote: "An independent American solution over British opposition was also more easily said than done. Aside from the damaging ill will it would create between our two countries, it would require the cooperation of the major American oil companies who alone, aside from Anglo-Iranian, had the tankers to move the oil in the volume necessary."

The Antitrust Division within the Justice Department resisted this course and instead developed a plan for participation by American independent companies, that is, companies which did not then have substantial interests in the Mideast.

Nonetheless, the majors were contacted and agreed to take part in the Anglo-Iranian concession. In return for this cooperation, Truman in the final days of his administration (January 1953) took steps to limit the cartel case. He instructed Attorney General McGrannery to proceed against the oil companies through a civil suit rather than by a grand jury investigation. Truman's memorandum said:

> Dear Mr. Attorney General:
> As a result of factors which have emerged since the institution of the current grand jury investigation of the international activities of the major oil companies, I am of the opinion that the interest of national security might be best served at this time by resolving the important questions of law and policy involved in that investigation in the context of civil litiga-

tion rather than in the context of a criminal proceeding. However, I believe that this would be the case only if the companies involved agreed to the production of documentary material which the companies are required to produce under an existing order of court based on grand jury subpoenas.

Consequently, I ask that you promptly confer with representatives of the companies to ascertain if they will agree to enter into a stipulation to that effect. If they will, I ask that a civil proceeding be instituted accordingly and that appropriate steps be taken to cause the termination of the pending grand jury proceedings.

Thereupon the Justice Department terminated its grand jury investigation and instead filed a milder civil complaint on April 14, 1953. This complaint named five defendants: Standard Oil of New Jersey (Exxon), Socony-Vacuum Oil Company (now Mobil), Standard Oil of California, the Texas Company (now Texaco), and the Gulf Oil Corporation. They were accused of monopolizing foreign production; dividing foreign markets; fixing prices worldwide, to the exclusion of competition; and monopolizing refinery patents.

Shortly after Eisenhower took office, he appointed Herbert Hoover, Jr., as his special representative to deal with the Iranian problem. Against the desire of the Antitrust Division, Hoover proceeded on the assumption that the only solution to the Iranian problem lay in utilization of the major oil companies as a vehicle for restoring Iranian petroleum production. On August 6, 1953, in a National Security Council Action Memorandum, Eisenhower designated the Attorney General to "develop a solution which would protect the interests of the free world in the Near East as a vital source of petroleum supplies," including development of "new or alternative legal relationships between the oil companies of the Western nations and the nations of the Near East." In resolving the problem, "it will be assumed that the enforcement of the antitrust laws of the United States against the Western oil companies operating in the Near East may be deemed secondary to the national security interest...."

Events in the Middle East moved swiftly to a conclusion. Allen Dulles, then Director of the Central Intelligence Agency, turned up in Switzerland for meetings with Loy Henderson, the Ambassador to Iran, and the sister of the Shah. Following that, American intelligence agents appeared in Teheran. The Shah dismissed Mossadegh, who predictably paid him no heed and remained in office. The Shah thereupon left the country. On August 18, units of off-duty police and military joined mobs in overthrowing Mossadegh. The Shah returned from exile to take charge of the country.

After the Shah was restored to power, Herbert Hoover, Jr., Eisenhower's oil emissary, set to work on reorganizing the Iranian oil industry.

In October 1953 the authority to direct a solution to the oil case was transferred from the Attorney General to the Secretary of State. On the surface the major companies showed little interest in Iran but were encouraged because of the downgrading of the cartel case to a civil action. They said they would be willing to participate in achieving a solution because of the "large national security interests involved." Negotiations between Anglo-Iranian and the American companies began in London, and the U.S. government stayed on the sidelines.

By January 1953 the broad outlines of the Iranian consortium had been discussed in the Justice Department. The objections of the Antitrust Division were noted, but the matter was out of its hands. On January 14 the National Security Council decided that "the security interests of the United States require the United States petroleum companies to participate in an international consortium to contract with the government of Iran, within the area of the former AIOC concession, for the production and refining of petroleum...." Thereupon the Attorney General gave his official blessings to the arrangement, ruling that the proposed consortium plan would not in itself constitute an unreasonable restraint of trade. On receiving this official clearance, the companies negotiated in earnest and reached an agreement called the Participants Agreement.

The details of this agreement were made public only in the spring of 1974 by the Senate Foreign Relations Subcommittee on Multinational Corporations. One provision permits companies with an aggregate of 31 percent or more of the equity in the consortium to set crude production at any level they choose, so long as it is below the level desired by the rest of the consortium members.

The Antitrust Division again protested this provision because it manifested "a continuation of the cartel pattern." But the Attorney General believed he had already "crossed this bridge," and approved the plan. More important, he "agreed that approval of the consortium was inconsistent with the cartel case as the complaint is drawn and that necessarily the case must proceed with emphasis on marketing aspects and not on the production control aspects."

The cartel trial never took place. Instead consent decrees dealing with marketing were entered against Exxon, Gulf, and Texaco. The case against Mobil and Standard Oil Co. of California were dismissed.

All these arrangements were made in secret, without the knowledge of the Congress and without the authority of law. (The only way the executive branch can give antitrust immunity to private corporations is under the provisions of the Defense Production Act.)

Inclusion of the major American oil companies in the Iranian oil industry killed two birds with one stone. It ended the cartel case and it perpetuated the cartel, extending its scope and making it an instrument of Cold War policy.

6. Coal

Coal has been commercially mined in the United States since 1750, when an English company opened a bituminous mine near Richmond, Virginia. Nine years later another bituminous mine opened in western Pennsylvania, and in 1791 anthracite coal was discovered in eastern Pennsylvania.

For a time the hard anthracite was believed to be useless, but its high-heat and long-burning qualities soon made it popular with local blacksmiths. Though its retail cost was 90 cents a bushel, anthracite from Lehigh, Pennsylvania, was advertised as a better value than the 2½-cents-a-bushel Virginia bituminous coal. In 1818 the Lehigh Coal Company obtained a charter from the state, leased 10,000 acres, and organized the Lehigh Navigation Company to develop a nearby river for transporting its coal.

Pennsylvania anthracite remained a local fuel source until 1820, when its earliest shipments were recorded. From then until 1840 anthracite was the only coal of economic importance in the United States. It was frequently gasified in the nineteenth century to make illuminating gas. Coal gasification, a technology still being researched today, involves incomplete burning of coal to distill gas from it, leaving coke as the by-product.

By 1854 bituminous output exceeded that of anthracite, and its relative importance has increased steadily since then. There are no important U.S. anthracite deposits outside eastern Pennsylvania, and today anthracite composes less than 3 percent of U.S. coal production, being used mainly in home heating.

Deposits of soft bituminous coal, however, were plentiful; during the nineteenth century thousands of mines sprang up in several states. By 1950 it was produced in Alabama, Arkansas, Alaska, California, Colorado, Idaho, Kansas, Kentucky, Maryland, Michigan, Missouri, Montana, New Mexico, North Dakota, South Dakota, Oklahoma, Pennsylvania, Tennessee, Texas, Utah, Virginia, Washington, Wyoming, Georgia, and North Carolina. In fact, bituminous deposits are so widespread that one nineteenth-century operator complained that it was impossible to obtain a monopoly on coal, since every time he bought out one mine, another opened nearby. Similarly, in his 1914 testimony before the U.S. Senate, mine operator Thomas McDonald bemoaned "the evils inherent in a wide distribution of ownership of coal lands. It results in such keen competition that the profits of the operator are reduced to a minimum."

Anthracite operators did not have this problem. Pennsylvania anthracite is confined to a limited area and was soon monopolized by the railroads. By 1887 the Delaware and Hudson Canal Company had invested over $5 million in anthracite mining, earning a 100 percent profit on its money. As early as 1887, this company was mining 13 percent of the total anthracite production. In 1920, nine railroads dominated anthracite coal: the Delaware, Lackawanna and Western; the Delaware and Hudson; the Lehigh Valley; the Reading; the Erie; the New York, Ontario and Western; the Lehigh and New England; the Central Railroad of New Jersey; and the Pennsylvania Railroad. By 1923 these companies owned 75 to 80 percent of all anthracite mined and 90 percent of all future supplies.

These rich anthracite fields were originally purchased from farmers for $50 to $100 an acre. The Philadelphia and Reading Railroad purchased 100,000 acres of anthracite land in Schuylkill, Pennsylvania for $40 an acre. Coal land buyers were so busy that by the mid-1920s there were no virgin coal lands left to sell.

Railroad companies purchased huge tracts of coal lands everywhere. Though they could not corner the bituminous market, they charged such high transportation rates that they profited greatly even on the coal they didn't own. (Even today 40 percent of the retail price of coal is absorbed by transportation costs.) Though U.S. mine operators usually fared better than their British counterparts, most of the profits from coal mining went to investors, landowners, and transportation companies.

In the early years of coal development, mining was a casual operation. Small mines were established almost as soon as an area was settled; these mines were usually operated by local people. Before 1885 almost all mining was done by hand, and mines were of very small capacity. In 1880, 2,944 bituminous mines produced an average of 14,269 tons each a year—the largest producing 332,056. Only 53 individuals or companies operated 2 mines each, and half of these mines were extremely small; 8 companies operated 3 mines each, 4 companies owned 4 mines, 3 owned 5, 1 owned 7, and 1 owned 9 mines—the largest 2 in western Pennsylvania.

When they ran short of wood, some farmers would dig up coal from their land to heat their houses. Many mines were one-man operations, by a prospector-miner. Sometimes operators didn't even bother to lease or buy their mines. Inhabitants of a few towns were granted the free use of nearby open coal seams, from which they would chip away enough coal to heat their houses. Had coal remained a local fuel source, as it was in the eighteenth and early nineteenth centuries, the tendency toward small owner-miners and community-owned deposits might have continued. But the plentiful supply of bituminous coal, coupled with eager investment capital, soon brought coal development to the point where production capacity outstripped local demand and new markets had to be found. Competition among mines was fierce. Transportation was scarce or expensive. Most operators were prospectors, not engineers; their mining techniques were inefficient and resulted in frequent accidents and low capital returns. Many companies failed, and the process of big mine owners buying out smaller ones was continual throughout the nineteenth and twentieth centuries. "But," says coal historian Howard Eavenson, "the constantly increasing demand prevented such cut-throat conditions as have existed in the industry since 1923." In 1920, 16 of Alabama's 100 coal companies went out of business and 24 more were absorbed by their competitors. By 1950, only 32 companies remained in Alabama—many of them connected with larger coal-using corporations (steel, railroads, electric power, by-products). In the late twenties, the frequency of mine failures increased due to large strikes, and even more went out of business during the Depression. In 1930 there were 1,355 idle mines in the United States—some of them shut down since 1923.

Before the Depression, there were no giants in the coal industry except in the anthracite region; no company mined more than 2.1 percent of the total bituminous coal produced. But by 1949, 1.9 percent of U.S. mines produced 29.3 percent of the country's coal. For the most part the companies that survived were owned either by the coal user, usually an iron or steel corporation (these were called "captive mines"), or by railroad companies. The small operator continued to be at the mercy of the railroads to get his coal to market. At the turn of the twentieth century, five-sixths of U.S. coal was sold—at prices lower than anywhere else in the world, and frequently below cost—to large transportation and manufacturing corporations.

A few fortunate mines in Pennsylvania, Ohio, and Indiana could ship their coal by canal. Several others were located near rivers and took advantage of inexpensive water transportation. Rail transport of coal was 40 times more expensive than water. But the canal era was ended through unscrupulous speculation and the advent of railroads before enough waterways were developed to free the coal industry from its dependence on the roads. By the end of the Civil War, most canals had passed into the control of their parallel railroads, which neglected to maintain the waterways. Soon canal transport became almost nonexistent. Had canals been built near more coalfields, the situation might have been different. But early canal development was not planned to provide coal transport, and only three canals were located near large coalfields. Canal builders were surprised to find that coal transport supplied the main income for these three canals.

Throughout the history of coal development, the availability of transportation was far more important than the quality of the ore or the ease of mining it. Eventually mines were opened only after a railroad line was built to them.

Coal's importance increased steadily during the first third of the twentieth century. In 1895 the value of coal represented 79.6 percent of all metals produced in the U.S.; from 1919 to 1936 the total value of coal exceeded that of all other metals. The wholesale price of soft coal rose from 99 cents a ton in 1890 to $4.88 a ton in 1949.

Since the supply of coal and the production capacity of U.S. mines always exceeded demand, coal was able to provide whatever amount of fuel the country needed, even during periods of great leaps in demand, as long as transportation was available. This plentiful supply was a vital ingredient in U.S. industrial development, as the growth of Pittsburgh, the first city built by coal, can testify.

Pittsburgh was planned around and grew on coal. In 1769, when Thomas Penn gave instructions for laying out the city, he said, "I would not engross all the coal hills, but rather lease the greater part to others, who may work them." As early as 1807, citizens complained that Pittsburgh's air was smoky and dirty as a result of coal burning in open grates. Pittsburgh's coal stimulated the birth of its steel industry; steel brought in other related industries; and all these relied on coal until the mid-1870s, when natural gas was introduced in Pittsburgh. By 1885 natural gas had displaced 50,000 tons of coal a year in Pittsburgh. (Within a few decades, this pattern was repeated in the rest of the country, as people began to replace coal heating with natural gas.)

Despite the many years when Pittsburgh coal sold at or below cost, by 1940 more wealth had been produced by the Pittsburgh coal field (which extends from Cumberland, Maryland, to Pittsburgh) than by any other single mineral deposit in the world—including diamond and gold mines.

Strip mining of coal began in 1914. Around 1920 it became important in the Midwest, though use of the technique was not significant in the East until World War II. One of the oldest methods of mining other ores, open-pit mining of coal was not established until large power shovels were developed.

Strip mining recovers 80 to 100 percent of the coal in a vein, compared with 40 to 60 percent for underground mining. The cost of stripped coal is over one-third less than that of underground-mined coal. A further incentive for companies to adopt strip mining was the fewer man-hours it required. In 1914 a strip-mine worker extracted 5.06 tons a day, compared with 3.71 tons for the underground miner. By 1949 a strip miner produced 15.33 tons a day, an underground miner 5.42. Given the economics of strip mining, it is no wonder the practice increased dramatically. In 1914, 35 strip mines using 48 power shovels and draglines produced 0.3 percent of all coal mined in the U.S. By 1949, 1,761 strip mines employing 29,267 men in all but five of the coal states produced 24.2 percent of the country's coal.

However, companies enthusiastic over the high profits they made from strip mining were not willing to expend these profits in land reclamation. The only successful reclamation during the first half of the century was done at government expense. The fourfold increase of strip mining between 1936 and 1946 led citizens, conservationists, and the United Mine Workers to join in an effort to stem stripping. As the outcry increased over the ravages of strip mining and the jobs lost by it, and as the cost of land reclamation—what little was done—continued to be laid to the public, state governments acted with restrictive legislation. But some courts overturned the legislation, claiming that the government had no right to require a landowner to "improve" his property. One law required companies to put up bonds, which would be returned to them after they reclaimed the stripped land. But in 1950, when the law went into effect, the strippers chose to forfeit their bonds rather than to go to the far greater expense of restoring the land. In effect, this law only kept the small operators, with little advance capital, out of the stripping business.

After Britain nationalized its coal industry in 1947, there was much discussion over whether nationalization was the solution to the problems of operators' profits and workers' satisfaction in the U.S. coal industry. The great fortunes from profits on coal capital had been made before 1885; since then competition and transportation costs had kept profits from rising too sharply. Operators compensated for this situation through lower labor and operating costs. The situation resulted in discontent all around.

The UMWA called for nationalization, feeling that the government would be a more sympathetic boss than the coal operators. The monopolistic control of anthracite mining, primarily by the Philadelphia and Reading Coal and Iron Company, led many others to support nationalization.

By 1947, 38 associations of coal operators had sprung up, partly to counter strikes and competition, and partly to lobby for their interests in federal and state governments. These groups fought strongly against nationalization. Ironically, what really defeated notions of nationalization was the very demon that these operators' associations were trying to exorcise—the continued existence of brisk competition among bituminous coal companies. The public felt that this competition was preferable to a government monopoly.

What the public did not realize was that the competition was in its last stages. There were almost no more small companies to gobble up. Already the giants were absorbing each other. In 1946 Consolidation Coal Company bought out Hanna Mining. Even separately, both of these companies were large enough to be among *Fortune's* "Top 500" U.S. corporations, and Consolidation was the largest coal company in the country. But less than 20 years later, Consolidation itself was bought out by Continental Oil Corporation.

Talk of nationalization eventually died out. But the federal government did take a more active role in the coal industry. The government first became interested in coal during the Civil War, when it was needed by the Union war effort. Congress levied a tax on the sale and consumption of coal, and for the first time had it valued in property tax assessments. (To avoid paying the higher land-tax rates, many companies began leasing their coal lands.) In more recent years the federal government has assisted the mining industry by funding research for more efficient mining techniques, uses of coal, and gasification technology; by funding surveys to find strippable coal and low-sulphur coal; by restoring strip-mined lands and leasing federal coal lands to mining companies; by conducting economic studies on how to expand U.S. coal exports; by taking over the unprofitable passenger part of the railroad business; and by passing legislation to ease miners' discontent.

7. Natural Gas

The first effort to transport natural gas for commercial use occurred in 1870, when a wooden pipeline was laid from West Bloomfield, New York, to Rochester, New York, a distance of 25 miles. The venture was a failure. In 1872 the first iron pipe was laid in the oil district near Titusville, Pennsylvania, then the main producing area of the country. During the middle 1870s gas was discovered in what is known as the midcontinent field—eastern and southeastern parts of Kansas. Since there were no sizeable markets nearby, it was necessary to transport the gas some distance to available markets. The availability of large amounts of gas and distant markets, then, led to the construction of long-distance pipelines. The first of these was built by Indiana Natural Gas and Oil Company in 1891 from Indiana to Chicago, a distance of 120 miles. The development of high-carbon, thin-walled steel pipes with tight joints that prevented leakage, and equipment for recompressing gas in transit, made long-distance transportation practical.

Some important interstate gas transportation existed previous to 1900, affecting the states of Illinois, West Virginia, Ohio, and Pennsylvania. Interstate movement of gas out of Kansas began in 1906.

The business grew steadily. In 1921, 150 billion cubic feet were transported for resale across state lines. About 65 percent of that gas was produced in West Virginia and moved out to Ohio and Pennsylvania. The next largest movement of gas in 1921 was from Oklahoma to Kansas. Interstate shipments of gas from Texas amounted to less than 3 billion cubic feet, or 2 percent of the total national shipments.

In 1930, 380.6 billion cubic feet were reported moving in interstate commerce, with West Virginia still the largest exporter, closely followed by Louisiana and Texas, in that order.

Seven years later the order was reversed. Texas was first, followed by Louisiana and West Virginia. The total value of all gas produced that year was estimated to be $527.5 million.

Pipeline Control

As the natural gas industry grew, the principal focus was on control of the big interstate pipelines. In its inquiry, the Temporary National Economic Committee said:

> The only feasible transport for natural gas at present is by pipe lines and therefore it is apparent that control of the only method of making natural gas available for use where an economic demand exists for it, must be considered in connection with the natural resource itself. The trunk pipelines are the bottlenecks of this resource and industry, and so the simplest way to control production is to control the means of transportation. Control of transportation makes possible control of prices both at the source and at the consuming end. There is no escape.

Before the Second World War, the industry was highly concentrated, with Morgan and Rockefeller interests dominating the pipeline business. There were four major pipeline groups:

The Rockefeller interests were centered in the Standard Oil Company of New Jersey, comprising 4,843 miles. These interests included the Interstate Natural Gas Company, operating in Louisiana and Mississippi; the Colorado Interstate Gas Company in New Mexico and Colorado; the Mississippi River Fuel Company, which had lines in Louisiana, Arkansas, Missouri, and Illinois; and the Natural Gas Pipe Line Company, which ran pipes from the Texas Panhandle to Chicago. In addition, through control of the Standard Oil Company of California, the Rockefellers maintained a 50 percent interest in the Standard Pacific gas pipeline, which serviced California.

The Morgan interests included the United Gas Corporation and the Columbia Gas and Electric Company. Between them these companies controlled more than 18,000 miles of pipeline. United had pipes running from Louisiana through Mississippi and Alabama to Pensacola, Florida, and from Texas to Mexico. It also was the largest minority owner in the Mississippi River Fuel Company. (This company actually was managed by Standard Oil of New Jersey.) Columbia Gas and Electric controlled Panhandle Eastern Pipeline Company, which had a line from Texas to Indiana.

The Lone Star Gas Company, with an integrated system in Texas and Louisiana, was indirectly affiliated with Columbia Gas and Electric, and hence the Morgan interests, through important stockholdings by individuals prominent in Columbia Gas and Electric.

Cities Service Company interests operated various petroleum and natural gas fields, and transported and distributed gas under franchises. Cities Service had holdings in the Colorado Interstate Gas Company and the Natural Gas Pipeline Company of America.

The four major pipeline groups produced 37 percent of all natural gas and bought up another 37 percent. They controlled 80 percent of all pipelines laid east of the Rockies, and one or another of these four groups was into every major gas-use center of the country.

Regulation

From its inception the gas business was wasteful. Beginning in the early 1920s, states passed laws aimed at conservation. But because of the growth of the pipelines beyond state boundaries, there were demands for federal legislation. In the 1930s Roosevelt first investigated the business; then in 1938, largely at the behest of Sam Rayburn, Congress passed the Natural Gas Act. It was meant to supplement the state regulations and control the interstate transportation of gas. The Federal Power Commission, which was to assume authority for the act, was to control the price of gas flowing in interstate commerce.

From 1939 on, the commission was torn over whether, in its efforts to control gas prices, it should also control the rates at which producers sold the gas to the pipelines. At first the Commission did not claim authorization over gas production on the grounds that the Natural Gas Act did not give it control. But in 1946 the Supreme Court authorized the Commission to extend its control over gas producers as well as pipelines.

At this point, the gas producers began a campaign in Congress to win exemption from control. In 1949 in the Senate Lyndon Johnson, from Texas, and Robert Kerr, from Oklahoma, pushed hard for Kerr's bill to exempt producers. Kerr was a big gas producer. Much to the irritation of Johnson, President Truman opposed the bill. It was at this point that Johnson turned against Truman and out of spite opposed the President's nomination of Leland Olds for another term on the Federal Power Commission. Olds opposed the Kerr bill. Johnson redbaited Olds, attacking him on the Senate floor. The Senate, siding with Johnson, refused to endorse Olds.

In 1950 Kerr and Johnson redoubled their efforts in behalf of the gas producers. As a result, the Kerr bill easily passed the Senate and scraped through the House with a 2-vote margin. When Truman vetoed it, Congress could not override his veto.

In 1954 the Supreme Court handed down the Phillips decision, which held that the Commission had authority over producer prices. There were more bills to exempt gas producers. This time the efforts were led by Oren Harris and William Fulbright, both of Arkansas. In early 1956 Congress passed this legislation; but there was a scandal. Senator Case from South Dakota revealed that he had rejected a $2,500 campaign contribution from a Nebraska lawyer representing the Superior Oil Company. Because of the bribe attempt, President Eisenhower, who was in basic sympathy with the legislation, vetoed it. And the Supreme Court's ruling in the Phillips case remained in force.

8. Electric Utilities

Early U.S. hydroelectric development began in the late nineteenth century; in 1896 Congress empowered the Secretary of the Interior to reserve choice water sites on public lands for electrical generating plants. By the end of 1910, 1,450 acres of public domain land had been set aside as waterpower sites.

The first electric utilities served very limited areas—usually a small city or village. Around the turn of the century, as rapid technological development expanded the distances of electrical transmission capabilities, these locally owned utilities began to merge. The purpose of the mergers was centralized control, not more electricity. In fact, the larger, central generating plants used by the new absentee-owned utilities had less capacity than the sum of the premerger generators.

Several of these enlarged utilities were incorporated into utilities "systems." Controlling interest in several systems was in the hands of one or more holding companies, which in turn were controlled by one top holding company. Sometimes this pyramiding was nine layers deep. The complex stock manipulations involved in forming these financial structures were arranged by promoters and bankers; as usual, John P. Morgan was foremost among these.

The house of Morgan was involved in almost every major stock deal of the period. Morgan was responsible for a complete reorganization of companies and stocks in the railroad industry, bringing in large chunks of investment from his English banking contacts. His specialty was liquidating highly competitive or failing companies and regrouping them into larger, noncompeting, companies with new stock issues and fresh capitalization. A substantial portion of the new stock always went to Morgan, giving him control over virtually every phase of industrial and financial activity in the United States. Morgan had engineered the merger of the country's two major electric equipment companies into the General Electric Company, receiving a controlling interest in the new company for his efforts. In 1905 GE formed the largest of the utility-holding groups, the Electric Bond and Share Company. During the 1920s, this "power trust" grew stronger by obtaining "the choicest water sites, buying up municipal plants, building high tension lines . . . and exploiting earnings through holding companies and overcapitalization." Described by one U.S. Senator as the most "gigantic monopoly" in world history, "it dwarfs," he said, "the Standard Oil Company in magnitude," the power trust developed rapidly and in obscurity until 1925, when the *Chicago Herald and Examiner* launched a series of attacks on Morgan's "vast General Electric combine." The newspaper accused GE and Morgan of complete ownership of the Electric Bond and Share Company, which, it said, "has for over 15 years successfully financed and supervised public utility companies in the U.S."

In addition to stock control of the holding company, GE was shown to have interlocking directorates with all Electric Bond and Share subsidiaries. The empire exposed was mammoth. One Electric Bond and Share subsidiary alone, the American Electric Power Company, itself had 46 subsidiary companies. All of these companies purchased their equipment from General Electric at above-market prices. This practice, combined with expensive stock manipulations and interest payments to the holding companies, kept the operating expenses of the utilities high. But the higher their operating expenses, the greater their profit, since the federal government had guaranteed electric utilities a "fair return" of 7 to 8 percent on their investment. Whatever money the utilities spent—whether on advertising to convince municipalities to sell their publicly owned utilities, on legal fees to counter local resistance to rate hikes, on interest payments for newly watered stock issues of their holding companies, or on high construction and equipment costs—these expenditures were considered part of "investment," and the utilities could legally raise their rates to consumers to cover the cost of this investment plus 7- to 8-percent profit.

Under these circumstances it was profitable for the holding companies to own every service the utilities might purchase. So, sandwiched among the several layers of subsidiaries and holding companies were a finance company to effect exchanges and issue securities, an "improvement" company to plan and finance large construction jobs, a construction company to carry out all new construction and repairs, and a management company to provide engineering, accounting, legal, and management services. With this setup the controlling group could collect profits twice—through the excessive prices the utilities paid to the four types of subsidiary service companies and through the interest on these expenditures, which could be obtained by rate increases.

The holding-company structure would also include gas utilities, coal mines, railroads, street railways, and water, oil, ice, real estate, and telephone companies—all brought in to decrease competition and increase profits. If any of these operations were drained by the controlling interests or showed a loss, the operating utility could simply purchase the failing company and deduct its losses from utility profits—thus necessitating another rate hike to achieve the "fair return." Component operations that showed a high rate of growth (often through high fees paid to them by the operating utility) were placed under one of the subholding companies in order to siphon off the profits.

Through the device of pyramiding, a relatively small investment to gain controlling stock of the company at the top allowed control of all subsidiary companies. By pyramiding its investments, by selling stock outside at a 5 percent return (thus gaining the 2- to 3-percent difference in the profit utilities were allowed to make on the money from the stock), and by watering stock (for every dollar invested in the utilities industry, three dollars of securities were issued), controllers of the top holding company were able to earn 55 percent annually on their initial investment, while still maintaining a controlling interest in all subsidiary phases of the operation. A million-dollar investment in the top holding company controlled $100 million worth of investments in subsidiary companies and returned to the original investor over half a million dollars a year.

But these profits were increased even more by continually reestimating the value of the utility companies' initial investment. Even without new capital expenditures, this investment figure—and thus the allowable profit on it—could be considered higher as land prices increased, as the replacement expense of buildings and equipment rose, and as the value of stock held by the utility went up. (Though for tax-assessment purposes, the private utility reassessed downward the "depreciated" value of its property.) In addition, private utilities added to their profits by padding costs. Auditors, examining the books of the Alabama Power Company, found that while the company claimed to have spent $11 million building a dam, actual expenditures were only $8 million.

In 1925, when a few Senators and newspapers began an attack on the power trust, 94 percent of all U.S. generating capacity was owned by holding companies. The Federal Power Commission had applications from the GE group alone for 4,255,000 hp of additional water power on public lands. Four-fifths of all generating capacity was in the hands of 41 subsidiary companies, most of which were owned or controlled by General Electric, the Doherty-Morgan-Ryan Company, and Samuel Insull.

Insull began developing his midwestern electrical empire in 1907, when he formed the Commonwealth Edison Company in Chicago. By 1911 he had purchased five electric utilities in the area, merging them into the Public Service Company of Northern Illinois. He also owned utilities in California, Pennsylvania, and Louisiana. In 1912 Insull created the Middle West Utilities Company, a holding company for his many interests, including Commonwealth Edison, the Public Service Company, and the Peoples Gas Company. By the 1920s all utility combines except Insull's were controlled by investment bankers, and primarily by Morgan. Insull became wary that Morgan would secretly buy a controlling interest in Middle West Utilities, especially when large amounts of MWU stock began falling to a hidden investor. Had Morgan's power trust not been under critical scrutiny by the press and the Senate, he probably would have moved on Insull's interests; but in 1928 Insull learned that it was Cyrus Eaton, not Morgan, who had purchased more MWU stock than Insull held. Eaton's interest in utilities was obscure. He had a near monopoly on midwest farming machinery but could not expect to best Morgan in a battle over control of utilities. Possibly, Eaton was acting as a member of the Morgan empire, since he already held a large interest in Morgan's Detroit Edison.

To counter Eaton's move, Insull pyramided his stock, forming his own investment trust, Insull Utility. In 1930 Insull unwisely bought Eaton's holdings at above-market value, on the advice of a banker friend of Eaton and Morgan's. This overextension, in a depression period, destroyed Insull. He resigned from the 60-odd corporations he controlled, declared bankruptcy, and fled to Greece in an unsuccessful attempt to escape indictments for mail fraud, embezzlement, and violation of the Bankruptcy Act. Most of his interests were eventually taken over by the Morgan group.

Public exposure of the power trust impelled Congress into action. Some Senators proposed a national power system that would furnish electricity at cost to all citizens. Others suggested giving the Federal Power Commission greater control over the industry. Though the FPC did make public much previously unavailable information about private utilities, it could not touch their financial superstructure. Established in 1920, the FPC's main function was to distribute public lands to power companies—especially national forest land in the West, which contained half the potential western water power source. During its first year the FPC received over 100 applications for power privileges totalling 7.7 million hp. The agency also investigates possible future sources of water power; examines the books of licensees; acts as attorney and judge in proceedings regarding licensees, rates, service, or securities; and "cooperates with private and public agencies" in the development of power.

The industry spent $450,000 lobbying to defeat a bill calling for Senate investigation of the power trust. The bill failed, but the Federal Trade Commission began an inquiry into the financial practices of the industry. However, these efforts were all but useless. Morgan's power trust continued unchanged into the 1930s. In fact, because of its federally guaranteed profit and its monopoly ownership of a necessity, the electric utility industry fared far better than most during the depression. By the mid-1930s six "superfinance" giants controlled the top utility holding companies. Two of these giants, the United Corporation and Electric Bond and Share, were totally under Morgan's control, and Morgan was suspected of having large interests in the other four: the American Super Power Company, Lee and Company, the Koppers Company of Delaware, and the United Founders Corporation. These six companies extended into all 57 of the then existing utility systems. Most of the 57 systems provided no electricity; they existed simply as controlling systems on their subsidiary operating utilities. Twelve of the 57 sold over 50 percent of all U.S. electrical power, though none of the 12 was directly engaged in furnishing electricity, with no operating subsidiaries. Twenty-two of the 57 had cornered 88.5 percent of the market, but only 2 of the 22 (the Edison Electric Illuminating Company of Boston and the Southern California Edison Company) were directly engaged only in furnishing electricity. The Electric Bond and Share Company owned the 4 largest systems—the American Gas and Electric Company, the American Power and Light Company, the Electric Power and Light Company, and the National Power and Light Company—servicing 12 percent of the market with 9.1 billion kilowatt-hours to almost 2.5 million customers.

The financial and ownership structures of these systems were complex. One system, the Cities Service Company, had 183 subsidiaries: 43 direct subsidiaries, which controlled 78 of their own subsidiary companies; these 78 companies had 49 subsidiaries, which controlled 14 other companies. The Cities Service group supplied electricity, gas, water, transportation, oil, fuels, and appliances. Another system, Standard Power and Light, had 127 direct and indirect subsidiaries, which sold electricity, gas, water, transportation, ice, oil, real estate, and telephone service. Electric Bond and Share's control of southern natural gas and fuel, only a small sideline for Electric Bond and Share, was described in a report of the U.S. Temporary National Economic Committee: Electric Bond and Share owned 47.2 percent voting stock in the Electric Power and Light Corporation, which had a 51.3 percent holding in the United Gas Corporation, which owned 46.7 percent of the voting stock in the Mississippi River Fuel Company and, through several of its subsidiaries, controlled a natural gas system from Louisiana through Mississippi, Alabama through Florida, and across Texas into Mexico.

All 57 systems had strikingly large capitalizations: 4 exceeded $1 billion and 9 were capitalized at between $500 million and $1 billion. But, in spite of huge capitalization (and because of it), the profits to be made were enormous, as the history of the Public Service Corporation of New Jersey makes clear. Incorporated in 1903 as the first utility holding company, Public Service of New Jersey provided electricity, gas, and transportation to 95 percent of New Jersey's residents. A relatively small holding company in 1938, and itself controlled by Morgan financial interests, PSCNJ had 85,000 direct stockholders and controlled subsidiaries that had 104,000 stockholders. During its first 36 years of operation, PSCNJ's subsidiaries spent $415 million in new capital investment. While these utilities' books managed to show only a legal 7 or 8 percent profit on this investment, the holding company's *yearly* revenues were almost one-third of the total 36-year investment of its subsidiaries.

In a study of federal rate controls over a variety of industries, the Temporary National Economic Committee found that electric utility rates could be considerably lower and still meet the government "fair-return" guidelines. The committee criticized the utilities' practice of offering low rates to industrial users. Since there was no need to compete for the business of individual consumers, who were a guaranteed market, the bargain rates were only offered to encourage industrial development, and thus greater electricity consumption.

These exposures heralded an attack by Franklin Roosevelt's administration on the utility holding companies. It was a battle that had been developing since 1912, when as a New York State senator Roosevelt first espoused public power and fought for passage of the Bayne bill, which authorized statewide development and sale of hydroelectric power to municipalities. The battle eventually culminated in the formation of the Tennessee Valley Authority and passage of the Utility Holding Company Act, designed to break up the power trust.

For many years before the New Deal, supporters of municipally owned electric utilities had fought a losing battle against the power trust. One of the most significant defeats for municipal power occurred in the 1920s, when the city of Muscle Shoals, Alabama, tried to buy electric power from the federally owned dam in Muscle Shoals. The government had been selling this power to the Alabama Power Company, a GE subsidiary, for one-fifth of one cent per kilowatt-hour; Alabama Power charged its customers 10 cents a KWH for the power—a 500 percent markup. Alabama Power vigorously opposed the federal government's dealing directly with a locality on the grounds that this practice would "set a precedent." Despite the fact that the municipality was offering to pay the government twice as much as Alabama Power paid for Muscle Shoals electricity, the federal government declined the city's offer. Nevertheless, the government also rebuffed repeated attempts by the power trust to buy the Muscle Shoals Dam outright, and eventually it became a part of TVA.

When the Holding Company Act was passed, the industry at first refused to comply with it and later fought it in the courts. But eventually the Electric Bond and Share Company was forced to break up its holdings, though most observers correctly believed that its "sphere of influence would continue to a considerable extent."

Despite these setbacks, by the end of the 1930s the industry had won its major battle. It had defeated public power. In 1939 only 7 percent of electric distribution was municipally owned; less than 2,000 municipal plants remained, and only 15 of these served cities of 50,000 or more population.

This victory seems remarkable, considering the fact that municipally owned power was consistently cheaper than private rates, two-thirds to four-fifths of which went to investors as interest or dividend payments on overcapitalized stock. In 1927 the average KWH rate in the United States was 7.5 cents, compared to 1.85 cents per KWH in Ontario, where utilities were publicly owned. In 1928 the U.S. householder paid ten times more than production costs for electric power. Tacoma, Washington's, municipal power plant charged 1.03 cents per KWH, while the nearby private plant in Portland charged almost three times more. The Portland company claimed its higher rate was caused by taxes. (For years private utilities fought paying taxes, claiming that the tax-exempt status of the nonprofit public utilities was an unfair competitive advantage.) But supporters of public power pointed out that the difference between the rates charged in Tacoma and Portland amounted to a yearly income for the Portland plant of eight times all city, county, and state taxes paid by the entire electric utility industry in Washington.

In recent years the utilities industry has taken advantage of expensive new technologies to gain tax advantages. The *Minneapolis Star* reported on March 26, 1974, that the

Northern States Power Company paid no federal taxes for 1973 and received an $8 million refund on taxes paid in earlier years as a result of expenditures on its Prairie Island nuclear plant.

In the 1920s, many citizens in Washington, D.C., were appalled to learn they paid 700 percent more for electricity than citizens in Canada's capital city. A major cause of high electricity rates in the District of Columbia was rising land values and building costs. The utility based its profit rate on what it would cost to replace its plant, rather than on actual cost. In the Twenties building and land prices were 50 to 60 percent higher in the District of Columbia than they had been when the electric plant was built. In 1937 TVA electricity cost householders from two-fifths of one cent to three cents a KWH, while the national average rate was five cents a KWH.

But despite such obvious incentives for public power, scores of cities sold their municipal utilities to private interests. Sometimes they capitulated to offers they couldn't refuse: private systems could lose nothing by paying far more than a utility was worth, since profit was always a fixed percentage of expenditure. Often a private system would launch a long, vigorous, and expensive advertising campaign to convince citizens that their municipal utility was a burden that could be unloaded at a profit if they cast wise votes in an upcoming referendum.

Citizens and municipal officials naively believed they could control the private utility to which they sold. But again and again the courts overruled any local ordinance that interfered with the federal fair-return guarantee. The problem was made even more complex by the existence of large interstate utilities. In one early case the U.S. Supreme Court ruled that the Rhode Island Utilities Commission could not prescribe the rate, even within federal boundaries, at which a Rhode Island corporation sold electric power in Attleboro, Massachusetts.

By 1966, 2,034 publicly owned utilities represented only 8.6 percent of total electric generation, while 278 private utility systems generated 77 percent of U.S. electricity; 10 federal agencies and 939 rural electric cooperatives produced the remaining 14.4 percent. Of the 2,034 publicly owned utilities, 1,927 are municipally owned and 107 are Public Utility Districts or state and county systems. They are

UTILITY

	Public	Private	Rural Electric
No. Customers (%)	9,555,298 (14.1)	52,424,660 (77.5)	5,688,553 (8.4)
Millions of KWH Sales (%)	145,974 (14.8)	798,505 (80.6)	45,688 (4.6)
Investment (%)	$8,184,451 (12)	$49,274,533 (72.5)	$3,838,735 (5.7)
Revenues (%)	$2,452,956 (13)	$14,374,168 (78)	$1,021,917 (5.5)

(The above figures are for 1966.)

spread throughout every state except Hawaii and Montana. The 15 largest are in Los Angeles, California; San Juan, Puerto Rico; Seattle, Washington; Memphis, Tennessee; San Antonio, Texas; Sacramento, California; Jacksonville, Florida; Omaha, Nebraska; Nashville, Tennessee; Phoenix, Arizona; Chattanooga, Tennessee; Knoxville, Tennessee; Everett, Washington; and Tacoma, Washington. Nebraska, Iowa, Kansas, and Minnesota have over 100 publicly owned utilities each; Rhode Island has only 1, West Virginia has 2, and Connecticut 6.

The creators of TVA and other federal hydroelectric installations hoped that this federal power could be sold cheaply to publicly owned utilities and thus curb some of the monopolistic control of the private systems. But the expensive electrical distribution lines needed to carry federal power are owned by private companies, and they refuse to allow transmission of federal power to municipalities along their lines. So the cheap federal electricity often has to be sold directly to private systems, which distribute and resell it at a higher rate. In 1971 TVA sold almost 75 million KWH to six private companies; 21 private utilities purchased over 1 billion inexpensive KWHs from five other federal installations: the Missouri River Basin, the Southeastern Power Administration, the Southwestern Power Administration, the

Colorado River Storage Project, and the Parker-Davis Project.

There are almost 200 federal hydroelectric generating plants in 28 states (Alabama, Arizona, Arkansas, California, Colorado, District of Columbia, Florida, Georgia, Idaho, Illinois, Kentucky, Michigan, Missouri, Montana, Nebraska, Nevada, New Mexico, North Carolina, North Dakota, Oklahoma, Oregon, South Carolina, South Dakota, Tennessee, Texas, Utah, Virginia, Washington, Wyoming). Many of the largest of these provide electricity for private systems: The Hoover Dam power plant, with a 1,344,800-kilowatt capacity, is operated by the Southern California Edison Company; the Grand Valley, Colorado, plant is leased to the Public Service Company; Cove No. 3 in Oregon is installed at a Pacific Power and Light Company plant. The Hetch Hetchy Dam, built with public funds to provide electricity to San Francisco, must sell its power to the Pacific Gas and Electric Company, which controls all distribution lines in the area. The Bonneville Project, a major federal producer of electricity, has been more successful than most federal installations in getting its power directly to publicly owned sources. But even Bonneville sold, in 1962, only 43.4 percent of its power to local government agencies and rural electric cooperatives; the rest went to industry (36.5 percent), private utilities (10.8 percent) and federal agencies (9.3 percent).

Federal electric power is produced by a variety of administrative structures. Primary among these is TVA, a government corporation; the Bonneville Project, a marketing agency of the Interior Department; and the Rural Electrification Administration, a lending agency originally meant to provide low-interest loans to publicly owned utilities serving areas of 1,500 or less population. The lure of the REA's low-interest loans was too great for private utilities. They demanded participation in the scheme, claiming unfair competition. They easily got around the population requirements by gerrymandering and now share in the REA's benefits.

Created by Congress in 1933, within 30 years TVA had

developed its generating capacity to 13.8 million kilowatts —six percent of the nation's total. Its largest customer is the Atomic Energy Commission, which uses 36 percent of TVA's power generation—more than Maine, Massachusetts, Rhode Island, New Hampshire, and Vermont combined. Even without its AEC business, TVA would be the nation's third largest electrical enterprise, after Pacific Gas and Electric and Consolidated Edison of New York.

Many conservationists have attacked TVA's use of strip-mined coal to generate electricity. Other critics argue that TVA sells electricity too cheaply, thus creating unfair competition for private utilities and encouraging excessive use of electric power. TVA's operating costs are less than half those of comparable private systems; it sells approximately 45 million KWHs annually at one-third the cost of private power. Ultimately TVA electricity reaches almost one million people.

Despite federal attempts to promote public power and private competition, a 1955 Senate investigation revealed that the power trust was not destroyed; the holding-company technique of absentee control through engineering and accounting firms continued or had revived, and Wall Street still dominated the industry. "The pattern which emerges from our hearings," concludes the Senate report, "is one of fairly consistent abuse of the monopoly position which private utilities enjoy." The report accuses the utilities industry of several serious abuses: giving bribes to local and state politicians, blocking construction of federal generating stations, conspiring to prevent the sale of federal power to public and cooperative utilities, influencing the Department of the Interior to discriminate against cooperatives, abrogating a federal contract with Southwest Cooperative, constructing "spite lines," discriminatory rates aimed at bankrupting cooperatives, engaging in political activity to gain leases on or purchase of public and cooperative utilities, and collusion to increase cost estimates of engineering bids submitted to TVA.

The most detailed accusation was levelled at Dixon and

Yates, two holding-company systems that had combined and were dominated by Wall Street. The Bureau of the Budget and the AEC, "in collaboration with TVA's worst enemies, behind TVA's back," had engineered the Dixon-Yates Contract, an agreement that gave private interests increasing control over TVA's power supply and allowed the holding-company combine "to participate in decisions involving TVA's future expansion." The Senate report says that private power, the Bureau of the Budget, and the AEC managed "to muzzle the chairman of the TVA" during a Congressional inquiry into the Dixon-Yates Contract.

The Senate report goes on to describe "a community of great banks, insurance companies and investment trusts, with interlocking directorates" among each other and with a variety of utility holding companies, including the Electric Bond and Share Company and Middle South Utilities. Large blocks of voting stocks in these groups were found to be centralized in a few financial houses on Wall Street. Large institutional investors interlocked with Electric Bond and Share, Middle South Utilities, and the Southern Company were: the Lehman Corporation, General American Investors, and Tri-Continental (all three also interlocked with each other); Harvard University and State Street Investments (also interlocked with each other); and U.S. and Foreign Securities, the Blue Ridge Mutual Fund, the Affiliated Fund, and the Investors Mutual Fund. Financial institutions indirectly tied to holding companies were the Marine Midland Trust, New York Life Insurance, Chemical Bank and Trust, Prudential Insurance, the Provident Loan Society, Metropolitan Life Insurance, the Chase Manhattan Bank, the John Hancock Insurance Company, Guarantee Trust, the J. P. Morgan Company, and the Provident Institution for Savings.

The Senate report had only scratched the surface, and concentrated primarily on abuses in the East. It urged further investigation into abuses in the Pacific Northwest, including accusations that private utilities placed their own representatives on Public Utility District boards, built uneconomic lines into cooperative areas, cut rates to destroy cooperatives, and undermined public power in a variety of ways. The 1955 report urges action against this "third great merger movement to sweep the country," reminding us that "the first two ended in disaster."

But nothing was done beyond additional investigations. In 1957 the House Committee on Government Operations investigated five power companies in the Rocky Mountain area, all subsidiaries of Electric Bond and Share, and accused them of writing a deliberately inaccurate and unsigned booklet that was used to influence Interior Secretary McKay and his assistants. The booklet, said the committee report, was charged against the companies' advertising expenses and paid for by increased rates; since 1953 its recommendations have been effected by the Department of the Interior, serving "to dismantle and subvert the federal power program."

Three years later, in another investigation, the Rural Electric Cooperative Association found substantial overpricing in 38 major utility systems, creating a rate of profit often above the permissible limit. Total estimated overcharges (based on a 6 percent return plus 1 percent reserves) for 9 of the largest systems from 1956 to 1960 were: the Alabama Power Co., $62,669,000; Appalachian Power, $72,008,000; Commonwealth Edison, $87,574,000; Detroit Edison, $36,225,000; Georgia Power, $72,367,000; the Northern States Power Company, $38,151,000; Pacific Gas and Electric, $76,394,000; Public Service Electric and Gas Company, $68,848,000; and the Virginia Electric Power Company, $83,619,000. The Southern Nevada Company had the highest rate of return, ranging from 7.9 to 10 percent.

In 1974 the Senate Committee on Government Operations conducted more investigations into the electric utilities industry and revealed that 14 banks are among the top ten securities holders in 10 or more major electric utilities each. Seven of these banks, all in New York City, exert control over most of the utility industry and own controlling interests in all of the 16 major utility systems investigated by the committee. The "Big Seven" and the number of utility systems in which they hold controlling interests are the Chase Manhattan, 42 utility systems; the Morgan Guaranty Trust, 41; the Manufacturers Hanover, 31; the First National City Bank, 29; the Bank of New York, 15; and the U.S. Trust Company, 12. The other 7 banks are: the Chemical Bank, New York, 10 utility systems; the Continental Illinois Bank, 12; the Girard Trust, Pennsylvania, 11; the National Shawmut Bank of Massachusetts, 11; the New England Merchants, 19; the Northwestern National Bank of Minneapolis, 21; and the State Street Bank and Trust, Massachusetts, 21.

PART TWO
The Energy Crisis of 1973–74

9. The New Industry

The Mossadegh coup (in 1953) was a turning point, setting off renewed demands for nationalism, eventually leading the way to a reorganization of the worldwide energy industry, and, in part, setting into motion the forces that created the current energy crisis.

As Eisenhower's oil emissary, Hoover arranged for the Iranian oil industry to be controlled through a consortium of companies, including British, French, Dutch, and—for the first time in Iran's history—American concerns. But the new oil consortium foolishly blocked membership by other countries, most importantly Italy, which depended on imported fuel supplies. This so infuriated Enrico Mattei, head of the Italian state enterprise ENI, that he determined to break the Anglo-American oil cartel.

Mattei got his chance immediately after the Suez crisis in 1956. With feelings running high against the British and the French, he helped the Iranian parliament write a new petroleum law, which introduced the hitherto unheard-of idea of joint ventures (rights to oil previously had been granted companies on the basis of concessions). Italy and Iran then made such a venture. Italy put up the capital; Iranians got the jobs in the oil fields. If oil was found, then Italy got first crack at it, but Iran got 75 percent of the profits from the venture's operation.

The joint venture between Iran and Italy served as an early model for what is now called "participation." Saudi Arabia made a similar deal with the Japanese. Mattei's techniques were adopted and elaborated by the younger and more militant Arabs.

Meanwhile the industry was tying itself in knots. As Joseph Stork observes:

> The profits-per-barrel of low-cost Middle East oil exerted a tremendous pressure on the companies to produce and sell as much as they could in order to maximize profits, a pressure which was only partially offset by the need to avoid "wasteful" competition for markets. As late as 1957, the monopoly position of the majors was considered strong enough to enact a crude price hike of nearly 22 percent in the face of declining production costs in the Middle East. This only increased the pressure to produce more at this new, higher profit-per-barrel. This led to surplus crude production and increased imports into the U.S. markets directly by the international majors to some extent, but significantly by smaller, independent companies who purchased the new supplies f.o.b. in the Middle East. This increase led directly to the mandatory import quotas of 1959.
>
> The imposition of the quotas protected the high-priced U.S. market, but only increased the supply of crude now forced to seek markets in Europe and elsewhere. Since the overall increase in crude production was more in response to higher prices than increased demand, this led to a situation of oversupply in which the pressure to cut prices was increased further. The pressure on European prices was increased still further by the reemergence of the Soviet Union as a major supplier of crude oil.

But as production increased, it became clear that the 1957 price hike could not hold. In February 1959, and again in August 1960, the official posted prices in the Persian Gulf were reduced. By reducing the posted prices, the companies were able to reduce the taxes they paid the producing countries. This enraged the Arabs, and together with representatives from Venezuela and Iran, they met at Baghdad to work out some form of collective action in defense of their countries' economic interests. The result was the formation of the Organization of Petroleum Exporting Countries (OPEC).

Nationalism was a major factor in persuading the oil companies to diversify their holdings. On the one hand, the major companies spread their search for oil into Southeast Asia, in the shallow seas off Indonesia, near Indochina, and running down to Australia. They moved actively into Alaska and from there into the Canadian Arctic. They stepped up a campaign in the United States for increased drilling on the Outer Continental Shelf. Most important, they turned back into the North American continent and began to buy up other energy sources: they bought into the coal industry, took a major position in uranium, and branched out into nuclear energy. In this way the oil industry became the energy industry.

The Structure of the Oil Industry

The structure of the modern petroleum industry is a web of intricate relationships that link the major companies one to another in a cartel-like system, protected and encouraged by the federal government.

The government plays a crucial role in the operation of the modern oil industry, and it is a role of increasing importance. Between one-half and two-thirds of future fuels (coal, oil, natural gas, geothermal steam, uranium, and oil shale) are located in public domain territories that are administered by the Interior Department.

The government takes a generally passive role in administering these territories, and in practice generally responds to the interests of the major oil companies.

Much of the future oil and gas is located under the sea along the Outer-Continental Shelf of the United States. This territory is leased out to the companies by the Interior Department in increasing amounts.

The process is instructive: two agencies within Interior administer this territory. The Bureau of Land Management runs the actual lease sale, but the U.S. Geological Survey, which acts as a scientific advisor, supervises the exploration and operations.

A company wishing to lease territory generally approaches the U.S. Geological Survey and requests an exploratory permit, which is routinely given. The company then conducts investigations, usually seismic surveys, undertaking occasional shallow core drillings. It then asks the department to put the territory up for lease, and historically the department usually goes along with the request. Because the U.S. Geological Survey has limited manpower and a small budget, it seldom obtains firsthand information on the extent of the minerals in the projected lease-sale area. Instead it purchases seismic data from the industry. The base data on this public resource are collected by industry and become a trade secret, and are not available to agencies administering the resource. With little knowledge of the mineral content of the area offered for sale, the Bureau of Land Management holds a bidding. The system requires a cash-bonus bid, meaning that the companies must advance cash before operating the lease. Often the bonus bid involved large amounts of money, running from several hundred million dollars into the billions. No small company, or even a coalition of small companies, can afford this bid. Inevitably the system results in a combination of large firms that bid on the property and then often operate it as a unit.

Virtually all the major companies are linked through bidding combines. Combinations for the purpose of bidding work against any real competition. Walter Mead, the economist, writes:

> In any given sale, it is obvious that when four firms ... each able to bid independently combine to submit a single bid, three interested, potential bidders have been eliminated; i.e., the combination has restrained trade. This situation does not differ materially from one of explicit collusion in which four firms meet in advance of a given sale and decide who among them should bid (which three should refrain from bidding) for specific leases, and instead of competing among

themselves attempt to rotate the winning bids. The principal difference is that explicit collusion is illegal.

Most of the major integrated petroleum companies hold joint interests with others in the transportation network that moves crude oil from producing regions to refineries and markets. As mentioned earlier, control of pipelines gave the Standard Oil Company its control over independent producers and refiners. And it is still true that independent producers must sell their supplies to those pipelines and independent refiners must buy supplies from the pipelines. As John Wilson, the former Federal Power Commission economist, points out:

> A notable difference between the past and the present is that today's pipelines are jointly owned by the integrated majors. Whereas early in this century Standard's control of the pipeline network gave it a distinct upper hand over all of its rivals, today's joint venture arrangements, which dominate the oil pipeline industry, draw ostensibly independent firms together into the common pursuit of a mutual purpose. Moreover, these jointly owned transportation links between producing, refining, and marketing operations (about three-fourths of all crude and one-fourth of refined products move through pipelines) require that producing and processing operations of the various partners be coordinated with each other so that the whole vertically integrated system functions smoothly.

While the major companies jointly bid on federal leases, they also join in ventures that produce oil through combined operations. Wilson goes on to point out:

> Only four of the 16 largest majors with interests in federal offshore producing leases own 50 percent or more of their leases independently. Conversely, 10 of the 16 own 80 percent or more of their offshore properties jointly with each other. In addition, very few companies outside of the top 16

have any independent holdings at all. In addition to the top 16, 23 medium to large size producers were surveyed. Of these, only two held as much as 25 percent of their leases independently and 17 had no independently owned leases at all.

The major oil companies are also interconnected through financial institutions, mainly banks. Whether or not the banks actually exercise working control, they have a capacity for potential control through stockholdings and interlocking directors. For example, in 1968 the Morgan Guaranty Trust Company had employees serving as directors on the boards of Continental Oil, Cities Service, Atlantic Richfield, Belco Petroleum, Columbia Gas, The Louisiana Land and Exploration Company, and Texas Gulf Sulphur. Morgan also held substantial stock interests in the Texas Eastern Transmission Corporation and the Panhandle Eastern Pipeline Company, as well as several gas utilities.

Banks and their trust departments often own substantial amounts of the common stock of rival firms in the same industry. For example, in 1968 Boston's State Street Bank and Trust Company owned 15 percent of the common stock of Texas Oil and Gas, 6.3 percent of the common in Amerada Petroleum, 11.6 percent of the common in Zapata Offshore, 10.1 percent of Kerr McGee, 7.7 percent of Pennzoil, and 7.6 percent of Newmont.

The California Oil Cartel

California provides an example of how this system of joint ventures works out in practice. There the Joint Committee on Public Domain investigated crude oil pricing. The committee issued subpoenas, and on the basis of information provided, Assemblyman Kenneth Cory testified that "the staff believes that virtually every joint venture among any of the major oil companies operating in California is a violation of Section 7 of the Clayton Act. . . ."

The committee looked into the bidding procedures for a

major oil field off Long Beach and focused on a joint bidding arrangement called THUMS. As Cory explained:

> THUMS is an acronym for five of the major oil companies in California, namely Texaco, Exxon, Union, Mobil and Shell. When the Long Beach unit was let out for bid, only two bids were received on the largest increment of the field. These were submitted by the THUMS group and a joint venture of the Standard Oil Company of California and Arco. The winning bid was submitted by THUMS which is now the operator of the field. It is important to understand that each of the members of THUMS is hypothetically a competitor of the others. By winning this bid, the five companies jointly took control, with joint production policies, of production from the Long Beach unit . . . [which] at one point totaled 20 percent of the entire production of the state and is now down to about 13 percent.

He continues:

> As is often the case when oil companies get together, the actual facts are stronger than the simple joint venture. We have received a series of documents pursuant to our subpoena which indicate until a few days ago prior to the actual bidding on the contract, an agreement had already been reached by Texaco, Exxon, Union and Mobil alone. They were prepared to submit a joint bid for the 80 percent increment of the Long Beach unit. Each of them had agreed to take 25 percent of their share of the Long Beach unit's production. At the last moment, Shell joined THUMS' group and each of the companies reduced its share to 20 percent. Shell was thereby eliminated as a potential competitor for the bid on the field. . . . To make matters worse, we have been told that in the initial discussions on the contract with the State, the State planned to require a "front money" bonus of $1 million. However, at the insistence of the majors, it raised this to $10 million, effectively precluding independents from bidding on the contract.

In a deposition to the Cory Committee, Otto Miller, the retiring chairman of the Standard Oil Company of California, described how joint ventures are arranged: Companies interested in a joint venture get together, go over geological studies, and haggle over what price to bid. Miller was asked, "And after discussions among representatives of the potential participating companies as to what the bid ought to be, if agreement cannot be reached, then the companies who have sent representatives to these discussions are free to go back and bid any price they want?"

Miller: "No. They are not. . . . Let's say there are two bids, one 60 and one 30. The person who wants to bid the 30, he can bid less than 30, but he can't go out and bid 70 and have it 100 percent."

"Why not?" asked the committee counsel.

Miller: "Because you had an agreement that we are going to try to reach a joint venture and you have agreed with this man, I will not meet your bid. See, he knows what this fellow is going to bid."

"Is that a written agreement?"

"It's an understanding," Miller replied. "I'm quite sure it's written."

Thus, the meeting prior to forming a joint venture becomes the point at which the industry is organized.

There are other interconnections in California. There are no common-carrier pipelines for crude oil. The majors own their own pipelines as private carriers.

> By state law, the private carriers must carry only the crude oil of the owner of the lines. However, the majors have circumvented this regulation by the creation of an exchange marketplace. When one of the majors wishes to move oil from a field in which he has no pipeline system to his refinery, he calls on another, which was a convenient pipeline, to "purchase" the oil in the field and sell back an equal quantity of oil at a point near to the first company's refinery. Thus the pipeline owner owns the oil only for the purpose of the trip through the pipeline. This has led to the creation of a massive

exchange marketplace, where oil is bartered. When dealing with each other, the majors do not buy oil, or sell oil, except with independents who are admitted to the pipeline system at the major's sufferance.

The pipeline system is also operated as a single, massive joint venture among the majors. The operation of this system itself leads to anticompetitive effects. For instance, we have been informed that when a refinery needs a particular type of crude oil to manufacture its required slate of products, the company negotiates with other majors to exchange a type of crude it does not need for the necessary type. The companies deal only with each other and not with the producer of the crude. One of the effects of this is that the majors do not go out into the field and try to acquire the crude they need by offering a higher price to the producer of the crude.

The Coal Industry

In certain major respects the creation of the energy industry was made possible by changes that occurred in the coal business, brought on largely because of the Cold War arms race.

During and after World War II, the coal industry was characterized by hundreds of little firms, locked into a boom-or-bust cycle, and engaged in furious battle with John L. Lewis' United Mine Workers. In 1950, however, Lewis made a settlement with George Love, then the leader of the coal industry, that set a long period of labor peace. Lewis agreed not to strike and consented to mechanization, which would cause a decline in the work force. After the settlement the union, under Lewis' leadership, actually bought into the coal business through stock purchases.

But more important, perhaps, than the union-industry agreement was the large-scale expansion of the Tennessee Valley Authority. Until the early 1950s the TVA produced electricity through hydropower projects. But in the mid-fifties, the demand for electricity was so intensified that TVA began to build coal-fired plants. The major purchasers of electricity were the plants that enriched uranium for hydrogen bombs. They were then, and are now, the single largest users of electricity in the United States. In order to supply the demands of the Atomic Energy Commission, the TVA created what were then thought to be innovative schemes for contracting out for coal. It established a long-term contract, guaranteeing firm orders for coal over long periods provided that the coal mines met certain levels of mechanism and other efficiencies. In doing so, TVA introduced the concept of the long-term contract, helping to drive out the smaller, less efficient coal companies.

The policy had certain ironic aspects. When the military ceased to need uranium enrichment facilities for the purpose of making bombs, they were then faced with a curtailment of operations. Confronted by the problem of what to do with excess capacity, they then turned to peaceful uses for atomic energy and thereby laid the foundations for the nuclear power industry.

The pressure for mechanization helped to drive the small mines out of business, concentrating control in the hands of a smaller number of large coal companies.

These events then made possible the outright takeover of the coal industry by large petroleum companies. The movement into coal can be traced back to the Standard Oil Company of New Jersey's decision in 1965 to begin assembling coal lands in Illinois as well as in North Dakota, Montana, Wyoming, and Colorado. Carl Herrington, a vice-president for Exxon, later explained to Congress why the company decided to become involved in the coal business:

> Our studies of the nation's energy requirements indicated that utilization of all the nation's energy resources would be needed to meet increasing demands. We concluded that coal mining and marketing of coal as a utility fuel offered an attractive long-term investment opportunity which draws upon Humble's experience in exploring for minerals and its established management and technical resources. Humble recognized concurrently that coal at some future date could become

a suitable raw material to supplement crude oil and natural gas and an economically attractive source of hydrocarbons.

Continental Oil Co., the twelfth largest oil company, purchased Consolidation Coal Co., in the fall of 1965. Consolidation was the largest coal producer in the country.

Howard Hardesty, at the time an executive vice-president of Consolidation Coal Company and a key figure in arranging the merger, later told a congressional committee the reasons for the purchase: "Continental acquired Consolidation Coal Co., I think for a whole series of reasons. The most important of which I think would be the fact that it recognized at the time that the supplying of energy to this nation just would not turn around solely on oil or gas, but would certainly rely to an ever increasing degree on coal and coal conversion to synthetic gas and liquids."

Other mergers followed. The Kennecottt Copper Company purchased the second largest producer, the Peabody Coal Company. Occidental Petroleum bought Island Creek Coal, the third largest producer, in 1968; the Standard Oil Company of Ohio bought out Old Ben Coal Company and Enos Coal in August 1968.

By 1970 the single largest block of coal reserves was controlled by Exxon, and two of the largest three producers had been purchased by oil companies.

By 1970, seven of the 15 leading coal producers were oil companies, including five of the largest. In the previous five years, oil companies had increased their share of the national coal production from seven to 28 percent. In a memorandum prepared for the American Public Power Association, S. Robert Mitchell, a former Justice Department economist, suggested that these figures might be on the conservative side. "These companies either buy coal from the other producers for resale or act as brokers for other companies in the sale of their coal. This effectively increases the control of the large companies that engage in such purchasing and brokerage activities, since the coal that is involved is produced by smaller companies."

The move by major petroleum companies into the coal business was encouraged by the federal government, under President Johnson, in several different ways. Perhaps the most important factor was private tax rulings by the Internal Revenue Service, which made it possible for oil companies with accumulated cash to buy other mining companies and pay no taxes. At the same time, the Antitrust Division of the Justice Department, under Donald Turner, determined after a careful inquiry, that the lead merger—Continental with Consolidation—was not a detriment to competition.

The Coal Rush

The excitement in coal centered on the virgin coalfields of the West. Beneath the sparsely populated prairies of the northern Great Plains lies the Fort Union Formation, containing the world's most valuable coal reserves, some 1.5 trillion tons of coal close enough to the surface to be economically mined. By contrast, miners in Appalachia work on seams anywhere from 3 to 6 feet thick. These coal reserves amount to 40 percent of all U.S. reserves and 20 percent of the world's reserves. They are particularly attractive because the coal is low in sulphur content, enabling cities and towns in the Midwest to burn coal in boilers to produce electricity and still meet air pollution standards.

But tapping these immense new resources raises a series of complex questions. Because the seams are so thick and close to the surface, mining will entail open-pit operations, which may leave the plains pockmarked with craters. Prospects for reclamation seem remote: the topsoil is fragile and thin; the prairies receive little rainfall. The prospects of rolling a few inches of topsoil back over the bottom of a 250-foot-deep pit and then trying to plant grass on it are not very hopeful.

There are other environmental problems. The underground coal seams act as a conduit for fresh-water providing an aquifer system. No one can say what will happen once this aquifer is punctured.

Beyond these immediate environmental concerns are serious political and economic questions. The region is water-scarce. In the past, available supplies of water went to support agriculture and ranching. But future plans call for the water to be used for coal mining and processing. Coal mining will import new industrial developments, creating industrial cities populated by people who work the mines; gasification and petrochemical plants; and the industries, such as aluminum, that settle around a power source and that in all probability will lead to a decline in agriculture. It also means the diversion of capital investment and work from other regions to the western mountains. It would lead to a further decline of Appalachia and the eastern and middle western heartlands that depended on those coalfields.

There are still other considerations. Exploiting and processing coal to make gas can result, in the view of some ecologists, in a worsening energy crisis because more energy will actually be consumed in the production of the synthetic fuel than the synthetic fuel itself provides, thereby tying the nation into a deepening and apparently endless series of energy crises.

Speculation in coal leasing was encouraged by the Interior Department's leasing policies. Under the Mineral Leasing Act of 1920, anyone can prospect for coal on federal lands in blocks up to 5,000 acres for a total of 46,000 acres in any one state. The prospector must apply to the Interior Department, then pay $10 for each 5,000-acre block as well as 25 cents an acre rent for five years. In the sixth year the rent goes up to $1 an acre. If a prospector actually mines the land, the government receives a royalty of 17 cents a ton for the coal.

Under the press of speculation, leases granted by Interior doubled in 1970. The government had leased out 767,902 acres of public lands in the western states for coal mining. Oil companies control about 24 percent of the leases outright. There are 520 leases outstanding, but only 73 of them are actually producing coal. Coal production on public lands actually declined from a peak of 10 million tons in

1945, when 75,000 acres were under lease, to 7.5 million tons in 1969, when 725,000 acres were leased out.

The largest single leaseholder was the Burlington Northern Railroad, which had acquired a checkerboard strip of coal lands as part of its right of way when it built the railroad across the West.

In October 1971 a group of 35 utilities and the Bureau of Reclamation issued the North Central Power Study, which formally paved the way for developing coal and water in a 250,000-square-mile area encompassing eastern Montana, western Wyoming, and North and South Dakota. The plan proposed building 42 power plants: 21 in eastern Montana, 15 in Wyoming, 4 in North Dakota, and 1 each in South Dakota and Colorado. By 1980 they would produce 50,000 megawatts of power, and 200,000 megawatts by the turn of the century. It would flow east and west over 760-kilovolt transmission lines. The major source of power production would be centered in an area 70 miles long and 30 miles wide between Colstrip, Montana, and Gillette, Wyoming.

In his study of the area, Alvin M. Josephy, Jr., writes in *Audubon:*

> Analysis showed that coal requirements for the 50,000 megawatt level in 1980 would be 210 million tons a year, consuming 10 to 30 square miles of surface annually, or 350 to 1,050 square miles over the 35-year period which the study proposed for the life of the power plants. At the 200,000 megawatt level, the strip-mines would consume from 60,000 to 175,000 square miles of surface during the 35-year period. In addition, each coal gasification plant, producing 250 million cubic feet of gas per day, would use almost eight million tons of coal a year, eating up more land, as well as 8,000 to 33,000 acre-feet of water (estimates vary widely) and 500 megawatts of electric power.
>
> The astronomical figures continued. At the 50,000-megawatt level nearly three percent of the tri-state region would be strip-mined, an area of more than half the size of Rhode Island. The transmission lines would require approxi-

mately 8,015 miles of right of way, which, with one-mile-wide multiple-use corridors, would encompass a total of 4,500 square miles, approximately the size of Connecticut. Power losses over the network of lines would exceed 3,000 megawatts, greater than the present average peak demand requirements of Manhattan, and would raise a serious problem of ozone production.

In Montana, the center of the coal rush, the usual technique was for land speculators to whip up a speculative frenzy, and thereupon to assemble leases in blocks so that a strip mine could be started.

While the federal leasing program made possible outright speculation in mountain states coal, much of the surface land was held privately by ranchers or farmers. The coal was held variously by the federal government, the railroads, the Indians, and the states, while the surface lands were held under homestead laws. Thus, it was necessary to dislodge the ranchers and farmers and to gain water rights for the necessary works.

To gather up surface rights, coal companies masqueraded as dummies. One called itself Meadowlark Farms. In the Bull Mountains, where Consolidation Coal made its first major bid to set up strip mines, representatives called on Ann and Boyd Charter, who owned a substantial ranch, and as Charter recalls:

> He got in the house and he said this is the deal. He said "I've got a contract here I want you to sign." I said a contract? He said, "yes." I said well, what's the contract about? And he said, "I want you to sign this: We can come on your ranch and explore for coal, and we can build roads, and we'll reserve the right to go through a fence any time we come to one," and he said they would compensate me at the rate of $1 for the whole thing. So I told this guy, you must think I look stupid that I'd give you the run of my ranch for $1. I said you'd better take that contract and get out of here. I said the door swings out, just the same as it swung in. And he started arguing with me,

that, "Oh this isn't going to hurt you any," and all this, that and the other. I said yes, it's going to hurt me and I ain't going to sign that thing and you'd better get out of here.

The Consolidation Coal representative thereupon went to each of Charter's neighbors, telling each one that the others had all signed away their surface rights. But the ranchers got together and stood their ground. They formed the Northern Plains Resource Council and fought the coal companies.

Coal Gas

Historically, the oil industry has been concerned for fear that coal might be converted into a synthetic petroleum that would undercut oil in price. When the German chemical maker I.G. Farben discovered a way to change coal into a synthetic fuel, Farben and Jersey Standard arranged a cartel so that neither company would interfere with the other's markets. During the Second World War, the German Air Force was fueled with synthetic petroleum made from coal. Since the 1930s the Lurgi Company has designed plants that turn coal into a gas, and these plants have been constructed on a commercial scale in places such as South Africa, Scotland, and on the Continent.

It was this expectation that coal, the most abundant of the fossil fuels, would become the mother fuel for a synthetic fuel business that apparently was behind the large-scale move by oil companies into the coal business during the 1960s, but their ability to do so was closely dependent on the natural gas business.

In the Phillips decision of 1954, the Supreme Court ruled that the government, through the Federal Power Commission, must regulate the price paid for gas by the big interstate pipeline companies to the major oil companies that produce the bulk of the gas. That decision represents a substantial effort to regulate the oil industry. Much gas is discovered in the search for oil, and in order to properly regulate gas prices, the commission would need to inquire into the costs of the oil exploration. That in turn conceivably might lead to regulation of the entire oil industry.

Shortly after the Phillips decision was announced, the gas men warned of a short supply of natural gas because, they said, there would no longer be any economic incentive to drill. There was a move in Congress to deregulate the price of gas. In 1955 A. P. King, Jr., representing the Texas Independent Producers Royalty Owners Association, said, "The supply of natural gas available in years past—at least in large part—was priced artificially low at the well and will rapidly fall short of growing demand unless allowed to reach normal competitive levels." The Independent Petroleum Association of America declared, "Federal regulation of natural gas will inevitably result in diminishing supplies and higher cost of this essential and desirable fuel."

Between 1954 and 1961 the FPC did not attempt in any serious way to regulate the price of gas. Then, during the Kennedy administration, Joseph Swidler, Chairman of the FPC, worked out a system of price control based on average existing rates for the different gas-producing areas. The oil industry fought area pricing through the courts and finally lost in the Supreme Court in 1968. Again gas producers warned of an "energy crisis," and sure enough, by 1969 gas reserve figures published by the industry began to decline.

Since the federal government relies exclusively on the industry reports for estimates of gas reserves, it has little way of checking whether the reserve estimates are correct or not. In 1971 FPC economists noted what they thought to be discrepancies in the figures of the American Gas Association, the trade organization that draws up gas reserve statistics. The staff economists asked the commission for permission to make an independent inquiry into gas reserves. The industry vigorously opposed such a study, and the full commission eventually denied the staff permission to make one. Instead for cosmetic purposes the commission ordered a gas survey, in which FPC staff members checked the industry reserve accounts but did not inquire into the background data on which reserve estimates were based.

The FPC accepted the industry figures, agreed there was indeed a gas shortage and an "energy crisis," and said the price of gas should be made higher to offer additional incentives to companies that searched for new gas. Thus, the price at the wellhead for gas offshore Louisiana was raised from 18 1/2 cents per thousand cubic feet (MCF) to 26 cents per MCF. Recent statistics for the year 1972 suggest that the price increase had a distinct effect: the opposite of the intended one. More successful gas wells were drilled in 1972, but strangely, gas reserves continued to decline.

In June 1973 James T. Halverson, director of the Federal Trade Commission's Bureau of Competition, testified before Senator Hart's Antitrust Subcommittee that there was "serious under-reporting" of reserves by natural gas producers to the Federal Power Commission. He went on to say that the procedures for reporting reserves by the American Gas Association "could provide the vehicle for a conspiracy among the companies involved to under-report gas reserves."

While the Federal Trade Commission experts were reporting their findings to Senator Hart, the Federal Power Commission was busily granting yet another price rise to the oil producers, an increase that would lift the price from 26 to 45 cents per MCF. Both the Federal Power Commission and the White House additionally sought to pressure Congress and the courts to deregulate the price of gas. If successful, the price of natural gas probably would go a lot higher than 45 cents.

It is important to understand that by 1974 the Federal Power Commission's incentive program had more than doubled the price without any indication that the industry had discovered new supplies of gas.

As the price of natural gas has climbed, the oil companies have begun to take a more active interest in synthetic gas, made from naphtha or from coal. Once the price of natural gas reaches a high enough level, then synthetic gas can become profitable, and it will be introduced to ease the "energy crisis" by supplementing the supposedly dwindling natural gas supplies. This is slowly beginning to happen.

The development of synthetics was hastened by the El Paso Natural Gas Company's much-publicized scheme to import large quantities of Algerian liquefied natural gas (LNG) for use along the East Coast. This project has been approved by the Federal Power Commission; and while it represents a large and important deal in its own right, it was crucial to the oil industry for little-noticed reasons.

Liquefied natural gas costs much more than natural gas. The gas must first be piped from wells in Algeria, run through a processing plant where the gas is frozen, then loaded into tankers for the trip across the Atlantic. The liquefied gas is finally unloaded at East Coast ports and turned back into a gas so it can be intermingled with natural gas flowing through the pipelines.

Among the major questions to be decided in the El Paso case was whether gas companies that purchased the Algerian LNG could "roll in" its high price. The commission eventually gave its approval. That decision is viewed as a precedent for the proposed synthetic projects, laying the way for the industry to roll in or pass along the extremely high prices of these schemes to the consuming public.

The major synthetic projects involve changing coal into a gas that can be intermingled in the pipeline system with natural gas. The technology is complex, with certain problems unresolved, but the profitability from the companies' point of view seems to be clearly established. That is likely to mean that these coal gasification projects—some 200 of them have been planned in the mountain states—will go forward. Coal gasification requires large amounts of strip-mine coal, large quantities of water, and the construction of expensive plants. If it goes ahead, it means large-scale development of coal resources in the mountain states, an attendant decline in the existing agricultural economy, and most important, a tremendous strain on the already scarce water resources of the region.

10. Oil Shortages

Throughout the 1960s Johnson, then Nixon, encouraged the oil companies in the reorganization of their business. The government helped bring about concentration of the industry through tax incentives and a nonexistent antitrust program. It encouraged development of a new coal industry in the West by making available inexpensive coal leases on federal lands. It pursued the development of nuclear power, while refusing to develop alternative energy sources. It sought to decontrol the price of natural gas. Perhaps most important, it steadfastly refused to make an independent inquiry that would have proved or disproved the industry's estimates of the nation's energy resources.

These policies were reinforced within the Congress, where the administration's tax and antitrust policies were understood and approved. The Congress has been conducting hearings and investigations on resource problems routinely and monotonously since the early 1920s. The Paley Report of 1952 dealt extensively with mineral resources, and Senator Jackson has taken testimony in his Interior Committee on energy since the late 1960s.

The energy industry, of course, had openly advertised an energy crisis, brought on, it maintained, by dwindling supplies of natural gas and oil. But the industry insisted it could develop additional supplies if it were allowed to charge higher prices to cover costs of mining scarce resources in out-of-the-way places.

The basic question remained the same: What is the extent of U.S. mineral energy resources? Where are these minerals located, and what are the costs of recovery and use?

Since the U.S. government does not conduct its own investigations of fuel resources, but instead relies on industry reports, it had little alternative but to take the industry's word for it.

Even as the issues became clearer and opposition to higher prices developed, neither Congress nor the Nixon administration would verify the industry's reserve estimates, let alone authorize an independent investigation of those reserves.

As the industry predicted, a series of rolling shortages developed; natural gas in the early 1970s, fuel oil and gasoline in 1973, and then a widening shortage of all oil-based products in the fall of 1973. The situation was purportedly made worse because of the Arab boycott.

Because neither the federal government nor the Congress have ever made any sort of serious, systematic inquiry into the workings of the energy industry during this period of turmoil, it is difficult to know the precise extent of the actual shortages. But the energy crisis can be set into a general framework that provides a measure of explanation for the events of 1973 and 1974.

In the 20 years between 1950 and 1970, domestic production rose from 1.9 billion barrels to 3.5 billion barrels annually. At the same time, U.S. production declined from 52 percent of worldwide production to 21 percent. During this period the price of oil rose slowly to $3.18 a barrel by 1970. To protect U.S. production, the import program limited foreign imports to less than 15 percent of domestic production.

Historically the American oil market was governed through the prorationing laws of the states of Texas and Louisiana. It was known as the balance wheel effect. When supplies were tight, Texas and Louisiana would have sufficient reserves to increase production to fill the market. But in 1972, with domestic production in decline, the states reported to the federal government that they no longer could provide a balance wheel. At this point foreign oil import quotas were, in fact, suspended, and crude oil imports rose, as did domestic production. In 1973 there was a further increase in imports until they accounted for about one-third of all U.S. supplies.

In the midst of this situation, then, the Arab countries invoked their boycott. At that point only 6.4 percent of U.S. imports came from the Middle East. Most imports came from Venezuela, Canada, and other parts of the world. Even with cutbacks in production, the United States was importing 790,000 barrels a day from the Middle East, a 63 percent increase over the same period in 1972. Even after the Middle Eastern countries announced a total boycott of crude shipments to the United States in October 1973, imports of crude remained at about the same level between September and December.

The Senate Report

In a special report on the oil crisis the Senate Government Operations Permanent Investigations Subcommittee strongly suggests that the shortage was contrived. An analysis of company reports to the Texas Railroad Commission by the staff of the committee indicates that refinery capacity had actually declined in the months preceding the fuel oil shortages of 1972. The Senate report states: "Five of the ten largest refineries actually reduced their refining capacity utilization during the first four months of 1972 in comparison to 1971. For example, Humble reduced its capacity utilization by 4.5 percentage points, Texaco by 3.3, Arco by 13.5, and Sun by 8.5 percentage points below the first four months of 1971."

The report goes on to say: "The low runs of the largest refineries during the first four months of 1972 resulted in the drawing down of fuel oil and gasoline inventories and was the beginning of the shortage problem. With low inventories of fuel oil at the beginning of the heating oil season, it was necessary for the giants of the industry to maintain maximum production. Instead they reduced refinery runs in the fourth quarter to 93 percent of capacity from 95.5 percent in the third quarter."

In early 1972 the administration was urged to increase the importation of crude oil to avoid a shortage. In February of that year, Glenn Willis of Clark Oil wrote to William C. Truppner of the Office of Emergency Preparedness urging an increase in crude imports, warning that "a severe crude oil supply crunch" might be at hand. He said: "We find ... everyone is using their import license at an accelerated pace as compared to 1971 while crude oil stocks declined to historic lows and all the companies, both independent and major, indicate a supply deficit beginning sometime in the third quarter and extending for the balance of 1972. Since little, if any, excess producing capacity exists in the United States, only through an increase in imports can the deficit be met." Pleas from independent refiners continued, and in April the Oil Policy Committee met to decide on future imports. At that meeting, Peter Flanigan, Nixon's aide in charge of oil matters, "questioned that we wanted to raise the level [of oil imports] to such an extent that the states would move back into market demand prorationing." This advice, however, was not taken, and in May Nixon raised the import quota level.

Despite the increases there was continued discussion among oil refiners and within the administration over a possible fuel oil shortage in the winter of 1972. In meetings with the industry, the Office of Emergency Preparedness discovered another factor in the looming shortage. That was price. In August 1971, when Nixon imposed price controls, prices for gasoline were at high levels and prices for fuel oil were low. Now the companies argued that while they had the capacity and economic capability to produce sufficient supplies of fuel oil for the winter, there was no incentive. An OEP memorandum of a meeting with Humble in July 1972 reports that Humble said No. 2 fuel oil was unattractive to produce because of price constraints and that the low frozen price of heating oil worked against the production of adequate supplies. In September 1972 Nixon encouraged the oil industry to import more oil to avoid a possible shortage by allowing them to draw down 10 percent of 1973 quotas.

But the oil companies refused to use their advance allocations. The Office of Oil and Gas report indicates that only 35 percent of the advance allocations were utilized by December 31, 1972.

Meanwhile the fuel oil shortage of winter 1972 came on as anticipated. Government officials met with major refiners and tried to persuade them to refine more fuel oil. At one meeting in November, Cities Service said their company was making 1½ cents more per gallon from gasoline than it would from distillates." The memorandum of the meeting states: "While the company has the ability to switch [from gasoline to distillates] they do not plan to do so. The company [and the industry] has a market for all the gasoline they can produce."

At a November 3 meeting with Exxon, Exxon officials acknowledged that the oil industry would have to run at 93.9 percent of capacity the rest of the year to produce adequate distillate supplies, "but economic incentive to operate at these levels does not exist."

The committee reports a meeting between Stan Stiles, vice president of Shell, and Mr. Roberts of the Office of Emergency Preparedness:

> Citing a 2.8 cents a gallon spread between its gasoline and distillate prices, Shell spokesmen said the firm was losing $1 million in potential profits by trying to produce distillate to meet its customer needs.
>
> However, as to going before the Price Commission to formally petition for relaxation of fuel oil price controls, the Shell people did not wish to make the effort.
>
> Mr. Roberts said in his memorandum: "Neither Shell nor any other company to date has been willing to go through the public hearing required for a price increase at the expense of the adverse publicity which would be incurred and necessary public disclosure of confidential company cost informa-

tion. All would welcome such a move, but would prefer that someone else do it."

The Cartel in Action

Meanwhile consumers in the United States bore the brunt of international oil politics, as American-based oil companies attempted to tighten control over markets and at the same time penalize North African and Middle Eastern countries for their insistence on participation and higher returns on their own oil. Officials from one firm, the New England Petroleum Corporation, described to Senator Frank Church's multinational subcommittee how that company was squeezed by the cartel and the U.S. State Department at the height of the energy crisis.

The New England Petroleum Corporation is an outgrowth of several companies that were formed in approximately 1935 and amalgamated into New England in 1945. It does business all along the East Coast of the United States, from Florida to Maine, and in the eastern sections of Canada running from Montreal east to the Maritime Provinces. It has eight large ocean terminals and a refinery at Freeport in the Bahamas, which is owned jointly with the Standard Oil Company of California. Standard owns 35 percent. The company operates 30 ships, 25 of which are on long-term charter. The company primarily makes fuel oils and petrochemical feedstocks. It does not manufacture gasoline.

Standard was a supplier of crude, and New England was anxious to enlarge and change the nature of its supply in order to furnish the big utilities in New York and New Jersey with low-sulphur fuel oil. It therefore agreed to Standard's proposition to establish a refinery, with Standard furnishing the crude. The refinery went on line in 1970.

Standard and Texas developed sweet crude in Libya, and at first they sought to sell direct to New England customers. The customers refusing, they then approached New England directly, and a long-term contract was drawn up. The Italians agreed to build the refinery; and in order to persuade Standard to guarantee the cost of the refinery, the Italians also agreed to buy crude from Standard. That oil came from Standard's supply in Saudi Arabia, which was held through Aramco.

The oil for the refinery was actually purchased at Freeport in the Bahamas. Standard supplied the ships and charged a rate of 30 cents a barrel for transport. This turned out to be advantageous from Standard's point of view. At the same time New England signed the contract with Standard for oil transport at 30 cents, the company also signed one with British Petroleum, also from Libya. The transportation cost on that contract was 18 cents.

Thus New England became tightly involved with Standard. Standard was part owner of the refinery, provided not only the oil but the shipping, and underwrote the cost of the refinery.

In early September 1973 New England received a telegram stating that the Libyan government had seized 51 percent interest in the Standard Oil Company producing in Libya, that Standard was resisting this takeover, and that all deliveries of Libyan crude from Standard's supplying company would be suspended as of September 1.

Mr. Richard Manning, New England's Attorney, explained what happened next:

> We replied that we did not agree that they were entitled to take this position, that we felt, as we saw it, certainly the 49 percent that remained in their hands was oil that was available to them and therefore it was available to New England and that so far as the 51 percent was concerned we understood that that oil was made available to them, and we had offered in the past, as we did with our other suppliers, we had offered to pay whatever the participation increases happened to be. We had already agreed with SoCal that whatever it turned out to be, if they had to pay a higher price, we would pay it.
>
> At the same time, of course, in almost as great a disaster to us, they, in effect, pulled the ships, they took the ships away that had been transporting this Libyan oil from Libya to the

Bahamas. At that particular point in time the shipping market was at an all-time high in the history of the shipping business because as the Libyan production had been cut during the course of these negotiations, the demand for shipping had drastically increased. The replacement oil for Europe primarily was coming out of Saudi Arabia, and the voyage is several times as long from Saudi Arabia to Europe as from North Africa to Europe. So as the crude oil, the replacement crude oil came into the market it took a great deal more shipping and had the effect of forcing the shipping rates up.

At this point New England contracted for Libyan oil through the Libyan National Oil Company at substantially higher prices, prices they would have been willing to pay Standard if it had bought back the nationalized crude and sold it to New England.

"You see," Edward M. Carey, the president, explained, "we supplied the Public Service of New Jersey, Edison Co. in New York, the New England Gas & Electric System, the Philadelphia Electric Co. in New York, the New England Gas & Electric System, the Philadelphia Electric Co., Niagara Mohawk System, the Orange and Rockland Counties, and when you put them all together it is a total of about 20 million people, and I think that we owed our allegiance to the 20 million people and we had to keep that going."

The utility companies themselves did not pay the increased cost of the contract. Carey continued, "Just so you understand who was paying the bill. I don't want you to think the utility company was picking up the tab; they were not. The utility was paying us, but they all have automatic fuel clauses and this was being paid by the individual in the street; this was being passed on right on down to ultimately the fellow who is least able to pay the bill—the consumer."

In addition to the loss of the ships, New England also lost the 90-day credit extension previously provided by Standard to pay for the oil. Now it had to push the utilities to put up the money in front.

Having at first sought to get Standard off the hook in

Libya, with no luck, and having been forced to do a contract with the Libyan national company, New England was now struck head-on by the cartel. What happened next is set forth in the record of the committee:

>Jerome Levinson [committee counsel]: Why don't you tell us what took place?
>
>Manning: On the 13th of September I was in the New England offices. I was in the office of a Mr. Weinand, who is a vice president, and he received a call from Mr. Folmar of the Texas Co. who identified himself as being the chairman of the Texas Overseas Oil Co. and he told Mr. Weinand that New England had, according to their information, put a ship into Libya to lift oil which had been taken from the Amoseas Co. in which the Texas Co. had a 50 percent interest. That this oil still belonged to the Texas Co. and that if New England lifted this oil, that the Texas Co. would pursue all legal remedies to prevent the use of this oil by New England. ... Within a matter of minutes, or it seemed like minutes, perhaps a half hour, Mr. DeBaum, who is another of New England's vice presidents, received a telephone call from Mr. Narvis, whom I have previously identified as being the representative of SoCal.
>
>Carey: Vice president.
>
>Manning: Vice president—as the gentleman we deal with on a day-to-day basis on matters affecting our relationship with SoCal.
>
>As Mr. DeBaum described the conversation to me—and as he subsequently wrote in a memorandum I believe you have seen—the conversations were so nearly the same as if to believe that these two speakers, the Texas Co. representative and the SoCal representative, were reading from the same script. It was the same conversation: "We understand that you put a ship into Libya," incidentally named the *Nepco Courageous,* and this was the first cargo of oil lifted from the Libyan National Oil Co. and, as I say, these statements made over the telephone were subsequently made a matter of record by letters from both Texaco and from SoCal.

Within another approximate 30 minutes, another telephone call came in—this time to Mr. Weinand—from a Mr. Mau of the State Department. Mr. Mau's conversation was essentially the same as the previous two except instead of saying that legal actions would be taken to preserve the rights of these companies, we were told that the State Department felt this was the wrong thing for New England to do, that this would have repercussions in the Middle East, and that the State Department opposed this action.

[Levinson continued:] If you had not been able to meet your contract commitments to the utilities, you would not have been able presumably to meet the payments on the debt which had financed the refinery, and under the previous arrangements which you have described as to debt arrangements also presumably—and please correct me if I am wrong—the ownership of the refinery would then have devolved to SoCal. Is that correct?

Manning: "Over a period of time, as the payments due to the Italians' banks fell due, if the revenues had not been produced from the crude oil produced in the refinery, we would have had no other source of payment. We had other crude oil but not a sufficient quantity to continue operating at a profitable level, which would have meant we could not meet our financial obligations, and SoCal, through this arrangement, through the mortgage notes, would ultimately have moved into a greater and greater equity percentage in the refinery.

Here, then, is a pretty clear picture of the cartel in action, showing how the major companies work in relation to one another and how the State Department operates as part of the whole.

An Artificial Crisis

In an analysis of the available statistics, *Environment Magazine* made a strong argument that the whole energy crisis of 1973–74 was artificial and contrived:

Oil Shortages 99

For the four months of 1973 when the supposed oil shortage was at its height, imports of crude oil were 50 percent above the same period of 1972. For the entire year of 1973, imports of crude oil from all sources ranged from about 2.5 to 4.1 million barrels per day and over the year exceeded total imports for 1972 by about 44 percent.

Domestic production, by far our largest source of petroleum products, declined by about 2.8 percent from 1972 levels, but our 1973 total supply of crude, both domestic and imported, was 5.8 percent above [that for] the same period for 1972.

During the summer of 1973, when the gasoline shortage was first announced, there was no reduction in total reserve stocks, which were at about 200 million barrels, or one month's supply for the nation. As of January 4, 1974, with long gasoline lines at the height of the shortage, there were 209 million barrels.

At the same date, supplies of distillate fuel oil were up 28.5 percent over those for the same week in 1973. Stocks of kerosene-type jet fuel were up 15.9 percent over 1973. The only petroleum product with lower reserves on hand during the first week of 1974 was residual oil, the fuel used in electric generation, and it was down by only 4.4 percent from a year before and higher than in any other weeks of January 1973.

Environment adds this:

If demand for any particular petroleum product had suddenly increased beyond the ability of the industry to meet it, reserves would not be expected to remain so high. The annual growth rate in consumption of motor gasoline has been 5.2 percent (about 4.5 percent for private automobiles.) For the first eight months of 1973 the growth rate of motor gasoline was 4.2 percent. Motor gasoline has amounted to 42 to 46 percent of domestic refined production for several years, with imports and exports negligible. Heating oil consumption has been irregular, but generally the growth rate has been much lower in recent years than the growth in gasoline consumption.

The energy crisis worked to the advantage of the industry. It pushed ahead long-term goals of persuading the government to back expensive research-and-development programs; began a debate on a national program to underwrite the energy industry's development of synthetic fuels; added to the pressure for more drilling on the Outer Continental Shelf, particularly in the Atlantic; hastened the development of western coal; postponed air pollution abatement; and forced up the price of natural and synthetic gas.

The energy crisis also contributed to a restructuring of the petroleum industry. A report by the Joint Economic Committee points out: "In the meantime the marketing structure of the industry has changed. Especially in the gasoline field, many independent gasoline marketers were eliminated in mid-1973, including some selling under major brand names. Total payroll employment at gas stations declined by 30,000 (five percent) between May and August. Independent fuel oil and propane distributors have also been under intense competitive pressure."

And it was profitable. The same report states: "Higher oil prices and marketing consolidations have raised oil company profits to extremely high levels. After-tax profits of 21 major companies for 1973 ran $9.3 billion, which was 58 percent higher than 1972. Much higher profits are anticipated for 1974. A recent forecast estimated the industry's 1974 cash flow would *increase* by $16 billion, some of which would be taken in tax writeoffs, with the rest showing up as profits."

11. Feeble Reform

On the basis, then, of this intricate history, which meshed the interests represented in Congress with those of the different administrations and industry, it would seem unlikely to find new or different policies emerging from the energy crisis of 1973. And indeed there is a good deal to suggest that Congress, despite loud protests, continued reworking the industry's own basic program: formally hastening the development of new coalfields, increasing the drilling on the Outer Continental Shelf, and reorganizing the tax system, all to the end effect that the energy industries would emerge stronger and more centralized.

Within the Congress, Henry Jackson, Chairman of the Senate Interior Committee, occupied a central position. Indeed it was virtually impossible to pass any energy legislation, effect any policy change, without Jackson's approval. Research and development funds, the Alaskan pipeline, synthetic fuels, the creation of the Federal Energy Office, information legislation, federal chartering—all went through Jackson's office. Matters were made complex because Jackson was running for President and using the energy crisis as a campaign issue. The Senate Interior Committee, with a staff of 90 or more, was in effect a Presidential campaign staff and hardly in the position to conduct any sort of dispassionate or rational investigation into the energy question.

Jackson himself maintained an almost oblique position. He has a reputation as a conservationist. Since 1969 he has received both the Sierra Club's John Muir Award and the National Wildlife Federation Legislator of the Year Award.

He won passage of major conservation legislation establishing the Redwoods National Park, the North Cascades National Park, and the Wilderness and National Wild and Scenic Rivers Act. He passed legislation to protect the last of the nation's wild horses, fought off Wayne Aspinall's Central Arizona Water Project, and was among the first to rise in defense of the Florida Everglades. By far his most telling contribution to environmental legislation is the National Environmental Policy Act, which requires environmental impact statements.

Among Jackson's major conservation achievements, and a source of considerable pride within the entire environmental movement, is a law called the Land and Water Conservation Fund. This fund, established in 1965, is used by the Interior Department to purchase lands for national parks and other preserves. In 1968 Congress agreed to increase the fund to $200 million, stipulating that additional funds come from revenues generated by leasing the Outer Continental Shelf for oil and gas. In this way conservationists tied themselves to the idea that some land could be saved by plundering other natural resources. To put it another way, money received from leases off Santa Barbara, where the big oil spill took place, went to help save the Indiana dunes, to preserve the Chesapeake Bay area, and to buy more land for the Point Reyes Park north of San Francisco.

This act is symbolic of Jackson's political views. It knots together contradictory concepts—economic growth and environmental controls. Although oblique in his manner, Jackson's basic energy policies elaborate on this concept. Thus, he is more than willing to slap at big oil companies for their pricing policies, and even to harass them with threats of federal charters and demands for fuller disclosure. But he knows, as do the companies, that this talk has no serious bearing on the industry's structure, which increasingly depends on the central government's providing companies access to large quantities of natural resources at low prices. And this is Jackson's thrust.

Strip-mine Legislation

While outwardly the energy crisis was taken up with oil shortages, actually basic policy centered on the coal business, whose future became rationalized through Project Independence, the administration's plan to implement the industry's program for a new energy industry based to a large extent on synthetic fuels made from coal. Oil shale, nuclear power, and increased drilling on the Outer Continental Shelf also figure prominently in Project Independence. But because of the vast amounts, coal was central to the plan.

What the industry desired and the administration in effect proposed was a large research and development program, steadily increasing prices for natural and synthetic gases, guaranteed markets at established prices for synthetic fuels, abatement of air pollution standards to allow industrial development in the western coalfields, and if at all possible, pre-emptive, weak federal strip-mine legislation that could be used to override opposition at the state or local level.

As the big oil and mining companies obtained coal and water leases in the northern prairies, and as they began to construct mine-mouth power plants and open-pit mine operations, they were met with spirited resistance by a group of ranchers and environmentalists. This battle was first joined in eastern Montana through a group of ranchers called the Northern Plains Resource Council. Later the council made a coalition in Washington with the Environmental Policy Center, which had pulled together several different environmental groups to lobby for a strong strip-mine bill.

All during 1973 and into 1974, the real focus of the energy crisis was on the strip-mine legislation. In 1973 the Senate had voted 82 to 8 to pass its first strip-mine bill. The legislation did not prohibit strip mining but contained requirements for reclaiming the land that had been mined, including a provision for returning mined areas to their "approximate original contour." The coal industry opposed the reclamation provisions as unnecessary and too costly. But it was much more worried by a section sponsored by Montana Senator Mike Mansfield, the majority leader. The Mansfield amendment was a direct response to rancher protests and was inspired in part by the Northern Plains Resource Council. It would prevent strip mining on any lands where the surface was privately owned but where the federal government had retained title to the mineral rights.

Obtaining clear rights to mine coal in the West is a complicated business because owners of the mineral rights to the coal are often different from the surface holders. Some of the land is owned by the federal government, some by the states, some by the railroads, some by Indian tribes, some by private owners. During the early 1970s the companies swapped land back and forth, attempting to hook together large contiguous areas for strip mines. The Mansfield amendment threw a monkey wrench into those plans. The National Coal Association estimated that the provision would prevent production of 37.5 billion tons of coal in the West. "In effect, it would wipe out any new operations in the West," Edward Phelps, president of Peabody Coal, said at the time.

In the House Interior Committee the strip-mine legislation was fought on regional lines, with representatives from eastern states working with environmentalists for a stiff bill in hopes of deterring the coal companies from quitting their deep mines in Appalachia for abundant supplies of less costly strip-mine coal in the western mountains, in particular, the coal in the Fort Union Formation. Congressman John Sieberling offered an amendment that would levy a $2.50 tax per ton on all coal mined. The mine operator could then obtain a tax credit of up to 90 percent for costs incurred in reclamation, safety equipment, or payments made for black lung. Since these expenses are heavy in the deep mines of the East, the Sieberling amendment was designed to restore a balance between deep and strip mines, and consequently between eastern and western mining.

Sieberling explained it this way:

The basic idea is that when Congress passed the beefed up

mine safety legislation it added $1.50 a ton to the cost of deep mining coal. That's one of the things that's given an impetus to strip mining and caused a fall off of deep mining. Hundreds of deep mines have closed. Let's redress the balance by imposing a flat fee on all coal mining and then allow against that fee a credit for black lung, safety equipment and surface reclamation. That is important because only three percent of our reserves are strippable and the rest have to be obtained by deep mining.

As the battle over the strip-mine legislation continued, the Environmental Policy Center sought to tie together environmentalists with other interests representing regional economic viewpoints. Representatives from the center met with labor groups, pointing out to them prospects for shifting employment patterns away from the East and Middle West as heavy industry settled around a new coal industry based in the western mountains. The same arguments were made with Congressmen from Appalachia and New England. The West Virginia State Senate voted to support the Mansfield amendment.

But the United Mine Workers, which had put itself on record against strip mining from time to time, vacillated. Arnold Miller, the reform president, had come out strongly against "high-wall" mining, the sort of stripping that cuts off mountaintops in Appalachia and pours debris down into the hollows. But when it came to stripping in the West, the union was not so clear in its opposition. The union's economic strength is in the health and welfare fund. Money for that fund is contributed by the coal companies, 25 cents per ton of coal mined. While most of the UMW members work in the deep mines of the East and would stand to lose their jobs if the industry moved west, most of the actual coal is produced from strip mines, and hence the union's financial strength is tied to stripping.

Moreover, the union faced competition among other unions for strip-mine workers in the West. The UMW worried about losing these potential members to the Teamsters or the Operating Engineers.

Until the very end of the strip-mine fight in Congress, the union vacillated and sat on the side lines. Both the Mansfield and Sieberling amendments were killed in the House Interior Committee, and the bill that finally was debated on the House floor was pretty innocuous. Essentially it permits strip mining when an operator can demonstrate that the mining area can be reclaimed. But the standards for reclamation are general, and it is unclear how or when they would be applied. The House bill includes short-term and long-term standards. The short-term, or interim, standards actually encouraged a strip-mining binge both in Appalachia and in the West. Over the long term, standards were more exacting in Appalachia, banning high-wall strip mining, but fairly lax for the West, where they would do little more than require operators to provide more information about their operations.

Most important, the bill does not prohibit strip mining even where, as in the West, the mining would ruin freshwater aquifers. Indeed it contains no standards for water at all. The bill banned stripping only in areas where reclamation is not "economically or physically feasible." But it exempts these areas if an operator has already made a substantial legal and financial commitment. On the other hand, the legislation can be viewed as the beginning of a strip-mine control program. In principle it places the burden of proof on the operator to demonstrate that reclamation can be achieved.*

Information

Disagreements over energy policy have often amounted to arguments about facts, that is, whether shortages are real or imagined, arguments over costs, and so on. Because of the

*The strip-mine legislation eventually passed Congress but was vetoed by President Ford.

government's historic role as a passive agent in energy policy, the information on which that policy is based is developed and used by the industry, even when the oil, gas, or coal are in the public domain territories. And as already demonstrated, this information is spotty or actually contrived to advance the interests of the companies involved.

William Simon, the administrator of the Federal Energy Office, himself acknowledged the situation before a subcommittee of the Joint Economic Committee in January, 1974, when he declared: "Let me say right at the outset that there has never been in existence an adequate energy data system.... Today and in the years ahead we need better data on everything from reserves to refinery operations to inventories.... Data we can check, verify, and cross-check."

And in an analysis of the energy situation, the Joint Economic Committee report put it this way:

> The lack of accurate, well-analyzed data regarding energy sources and uses has placed the United States government in a ludicrous position. Even those officials directly charged with administering energy policy are unable to determine accurately the extent of the present fuel shortage or to estimate reliably its potential impact on the economy. Nor can they determine fuel production costs with anything approaching the degree of accuracy necessary to administer the price control program. The government knows almost nothing about the extent of the vast mineral fuel resources contained in public lands. Tax policy formulation is hampered by the lack of analysis of existing special tax provisions for mineral fuel extraction and consequent ignorance of their impact.

But while information was crucial to resolving the energy crisis one way or another, neither the administration nor members of Congress sought legislation to gather data independently on energy resources. The closest Congress came to deliberating such legislation was a bill sponsored by Gaylord Nelson, the Democratic Senator from Wisconsin. This bill, to establish a National Energy Information System, authorized the Department of the Interior to prepare an inventory of the nation's energy resources. The information system would include a public library of energy information for public use, a confidential library for restricted government use, and a secret library for use only in preparing anonymous statistics. The purpose was to gather all available energy information, but then to protect sensitive company information or matters of national security importance by placing them in the confidential or secret library sections.

All energy companies would be required to submit accurate information to this system. Failure to do so would bring fines and prison sentences. The director of the information system was given subpoena power to inspect books and records of the reporting companies.

Additionally, the Secretary of the Interior would be directed to compile and maintain current an inventory of all mineral fuel reserves and natural energy resources in the public lands of the United States, including the Outer Continental Shelf. The Secretary could verify reports of minerals on privately held lands by inspection.

The legislation required that all energy companies report full details of all mineral fuel reserves and natural energy resources that they control in any part of the world.

And it required that companies report on the basis of each establishment, that is, for each operation (oil well, pumping station, tank farm, etc.). This sort of reporting makes it difficult to disguise or fudge statistics.

The history of the Nelson bill is instructive: The Senator introduced an early version of his library concept in the summer of 1973 as an amendment to the Alaskan pipeline bill. In debate on the Senate floor, he agreed to withdraw the amendment when the pipeline leader, Senator Jackson, promised to join Nelson in sponsoring a more detailed version of the legislation, and to hold hearings jointly with him. Nelson has no important committees. Here Jackson was offering him the auspices of the Interior Committee, which he chairs. Nelson thereupon agreed to the arrangement. A bill was drawn up and hearings held. Jackson himself paid little attention to the legislation, but there was an understanding that Nelson's bill would be permitted to move through the Senate.

During the markup sessions on the bill, William Simon, the FEO administrator, and the industry opposed in principle the idea of making corporate information public. There were specific attacks on the sections that required reporting by "establishments" on the grounds that they compromised the companies' competitive position. As a result the Nelson forces bargained away the "establishment" reporting concept to keep the bill alive.

It then became known that Jackson had discussed the Nelson legislation with Harley Staggers, Chairman of the House Commerce Committee, the group that would handle the bill in the House of Representatives. Staggers reportedly told Jackson his committee would not even consider the bill. Staggers reasoned that all the powers provided by the legislation were already provided by the legislation that created the Federal Energy Office. The Nelson bill was redundant, in Stagger's view.

Nelson pushed on. Then in May, apparently at the suggestion of William Simon, Senator Mansfield sponsored a proposal to create a special task force that could study resource shortages and supplies. Mansfield's proposal represented the leadership position, and it meant Nelson's bill was finished. Nelson put up a fight on the floor but lost. After passing the Senate, Mansfield's proposal was lost in the House Banking Committee.

The Federal Energy Corporation

During the early part of 1974, the staff of the Senate Commerce Committee worked on a draft bill for serious, long-term energy reform that would formally bring the oil industry under regulation and create a federal energy corporation to operate as a competitor to private industry. These proposals were included in the Consumer Energy Act. The scheme was originally sponsored by Senator Adlai Stevenson III, of Illinois, and the Commerce Committee staff rounded up more than 30 Senators to cosponsor it. The bill was written, then rewritten with the help of a small group that included Ralph Nader; David Freeman of the Ford Energy Policy Project and a former energy adviser to Nixon; Lee White, the former Federal Power Commissioner, and representatives from the public power lobby and other liberal groups. Environmental lobby groups did not take much of an interest in this legislation. Most important, Senator Jackson was against it.

This proposal reflects the most progressive rendering of Neo-Populist thinking. An early explanation of the act provides a good explanation of that position:

> The principal thrust of Working Paper No. 1 of the Consumer Energy Act is to create incentives toward a gradual and orderly restructuring of the natural gas and oil industry by enlarging the market share and influence of the independent producer. Currently, there are more than 10,000 oil and gas producers in the United States. Approximately 90 percent of the Nation's production, however, is controlled by a dozen natural gas and oil companies. Because of increasing concentration, the ranks of the independent oilman have, until recently, been dwindling rapidly. In 1954, there were about 40,000 oil producers in the United States. Today, there are less than one-quarter that number. These independent producers are often small entrepreneurs who are willing to risk their fortunes on the discovery of oil. Although they do 80 to 90 percent of the wildcatting in the Nation, many of them cannot afford to engage in expensive production drilling, and as a result, most of the oil and gas production in the United States today is *not* done by the independents, the risk-takers, but by the major oil companies. As the cost of drilling increases, as seismic and geological techniques become more sophisticated, as new discoveries are made in more remote areas or far off-shore, as lease bonuses increase and as construction costs soar, the independent oilman is becoming an endangered species. The proposed bill is intended to reverse this trend.

It provides special incentives to independents—price deregulation, access to Federal lands, access to oil pipelines, protection for both independent and franchised dealers. It is hoped that with these incentives the market share of the independent sector of the oil industry will increase from the current 10 to 40 percent in a decade. Independents will then be a viable competitive force in the marketplace; at that time it may be appropriate to review the need for price controls on major producers.

The Consumer Energy Act of 1974 also proposes new incentives for the major oil companies. The premise of the bill is that the free market is the best allocator of scarce resources, but for the free market system to work properly, multinational oil companies must be structured competitively. Mounting evidence suggests that the discipline of the marketplace is not operating to establish reasonable oil prices for products sold by major petroleum companies. Instead of market forces, the world oil price is determined by an oil-producing cartel. Oil that is selling on the world market for $11.60 costs but $0.15 to produce in the Middle East. World oil prices and, consequently, domestic oil prices have soared not because of the operation of the free market but because of the monopoly power of a few Middle Eastern nations.

Since the discipline of the free market is not functioning effectively with respect to the major oil companies, the consumer is faced with two choices: (1) tolerate the distortions and misallocations resulting from oligopolistic or monopolistic supply and pricing patterns, or (2) endure the difficulties associated with regulation and effective control. The Consumer Energy Act proposes, for the major companies, a firm, fair, streamlined, and workable system of price controls and incentives to encourage maximum efficient levels of production of oil and gas at reasonable prices. This goal would be achieved by a reformed system of Federal Power Commission oversight and the establishment of a Federal Oil and Gas Corporation as a supplier of last resort.

The measure goes on to propose regulating well-head prices of both oil and natural gas "because natural gas and oil are generally produced by the same companies, often from the same wells, and because both fuels are generally end-use substitutes for each others." But it would deregulate small producers. Refinery rates would also be put under regulatory control.

The Federal Oil and Gas Corporation

> . . . is intended to satisfy national energy needs, stimulate competition in the petroleum business, and provide the public with knowledge of the actual cost of producing oil and gas so that public policy can be geared to the Nation's interests. It would give the Nation a yardstick against which to judge the performance and pricing of the private oil companies. It could provide a means for effective exploration of Federal lands, the establishment of strategic reserves, and provide the Nation with a supplier of last resort. It is not the purpose to provide a forerunner for nationalizing the American petroleum industry. Its purpose is to develop public resources while preserving the free enterprise system in the oil and gas industry. It would provide a spur, a yardstick, an incentive for competition.

The corporation would give preferential treatment in its operations to independent producers and would also give preference to states or political subdivisions within states. The directors would be appointed by the President, and to insure that the corporation would not become a threat to the existing oil industry, the act would limit the oil operations to 20 percent of the total public territories offered by the government for leasing.

Tax Policy

Since the 1920s the petroleum industry's major policies were effected through federal tax policies. The oil industry receives three different sorts of specialized tax treatment: depletion allowance, intangible drilling costs, and foreign tax credits.

Feeble Reform 113

The idea of depletion is that an oil well, like any other sort of business, wears out. The tax law allows companies to subtract from income, or depreciate, a suitable amount to cover the loss. In the case of oil, the original tax provisions were called cost depletion and were based on the cost of what the oil company actually lost. They were similar to depreciation provisions, which allow a business to subtract each year an amount to cover the wearing out of buildings or equipment.

But because it was difficult for the IRS to know how much oil was left in a well, and hence figure out how much to let a company deplete each year, Congress instead adopted in the 1920s a percentage depletion allowance. The argument by oil state members was that it would be easier for the IRS to administer and thereby prevent excesses of the oil laws. The depletion allowance was set at 27 1/2 percent and since has been reduced to 22 percent.

As it turned out, the depletion allowance had unanticipated results. Many of its benefits go to foreign operations or to individuals who do not actually produce oil. A landowner who receives royalties from an oil company gets the benefits of percentage depletion, even though he has nothing to do with exploring or drilling for oil. A 1968 Treasury Department study concluded that 42 percent of the depletion allowance goes to such nonoperating interests in both domestic and foreign operations.

According to *People and Taxes*, Ralph Nader's report on tax policies: "It [the depletion allowance] does not encourage them to explore for new oil, but rather to drill more holes in known oil fields. Only 10 percent of exploratory wells strike oil, so depletion benefits only one tenth of the really risky drilling. That's why oil companies prefer to spend their money drilling in existing oil fields where they know they will get the depletion tax subsidy."

In addition, because the allowance is based on income, as prices and profits skyrocket, so does the depletion allowance. Instead of paying more taxes on more income, the oil companies pay less. In 1972 the depletion allowance cut the oil industry's tax bill by 15 percent, costing the Treasury $1.7 billion.

A special tax deduction for intangible drilling costs allows oil companies to deduct most of the cost of drilling wells immediately instead of spreading out those deductions over the years the well operates. Intangibles include such costs as labor, supplies, and repairs. Such intangibles usually amount to 70 to 90 percent of the total cost of drilling. The intangibles cost the Treasury $650 million in 1972. The argument is that intangibles also are a prod to the oil companies to get out and explore for new oil, but the intangibles are available for wells in old fields. Why bother drilling in a new, uncertain area when you can produce oil, get the profit, and also reap the intangible write-off in an old field?

The foreign tax credit has been a major governmental policy, undertaken in 1950 to encourage the development of the international petroleum cartel as an instrument of foreign policy. This credit lets the oil companies subtract from U.S. taxes the taxes they pay in foreign countries. But as indicated earlier, these taxes paid abroad are not really taxes, but royalties paid the Arab countries for their oil. The amounts are much larger than ordinary taxes, sufficient to erase taxes oil companies might be charged on foreign income, and large enough to shelter other income from refinery operations, from shipping, or from dabbling elsewhere.

As *People and Taxes* observes:

> Tax 'incentives' for the oil industry just haven't worked. In fact, they have encouraged exactly what the American public does not need: Oil drilling and production abroad instead of at home; wasteful drilling in existing oil fields instead of exploration; monopoly control of the oil industry instead of competition; and a lack of energy sources to compete with oil. The tax incentives have helped no one except the oil industry. That is why we should get rid of them. All of them.

During the height of the energy crisis there was a move within the House Ways and Means Committee to rewrite oil tax policies. But it was a sleight of hand. While Wilbur Mills and other members agreed to a reform of eliminating the oil depletion allowance, they increased the amount of income that could be sheltered under the intangible drilling cost write-offs—in effect, simply changing the form of the tax gimmicks, not reforming them.

Solar Energy

Nowhere is the government intransigence on energy clearer than in solar energy. Solar energy involves a relatively simple process in wide use for many years. Heat from the sun is absorbed by collectors, often located on a building's roof, which in turn heat a fluid such as water. A collector might consist of a black metal plate with copper tubing soldered on the back. The front of the plate has a glass cover, and the whole apparatus is set into a kind of window frame. Water runs constantly through the copper pipe and is warmed by the sun; it then goes into a storage tank.

When supplemented by an alternative system for use when the sun is not shining, this device can provide hot water for household use. If insulation is good, the hot water can be used to run an absorption unit for a refrigerator or an air conditioner, or it can be used for space heating.

Solar energy can be an important factor in reducing energy use within a building. A building's energy use depends on how it is placed, whether for example, it is shaded, the windows open and shut, its walls are thick enough to retain heat and coolness, the materials it is made of, insulation, and so on. In this overall context solar energy becomes an important tool in reducing reliance on fossil fuels.

The application of solar energy is not new, though its use has generally been limited to heating water. In the United States solar water heaters were made from the turn of the century through the mid-1930s in California. During the 1950s thousands of solar water heaters were sold in Florida; but in both California and Florida solar heaters were discontinued because of the introduction of low-cost natural gas.

In Australia the government requires that new houses in the northwestern part of the country be equipped with solar water heaters supplemented by electrical back-up systems. In both Israel and Japan solar water heaters have been used widely. And in the United States several dozen buildings are heated by the sun's rays. The U.S. government, through the General Services Administration, is building two energy-conservation solar office buildings—one in Manchester, New Hampshire, and the other in Saginaw, Michigan. The National Science Foundation was spending over $12 million on solar research in 1973, and there were several big office buildings announced, as well as a variety of projects by architect-builders.

In December 1972 a special report by the National Science Foundation and NASA concluded: "Solar energy is received in sufficient quantity to make a major contribution to the future US heat and power requirements.... There are no technical barriers to wide applications of solar energy to meet US needs." The report predicted that solar energy might economically provide up to 35 percent of total building heating and cooling loads—or, looked at differently, 20 percent of the nation's electric energy requirement and 30 percent of its gaseous fuel needs. If successful programs were established, the report says, "building heating systems could reach public use in five years, cooling systems in six to eight years, and electricity production in ten to fifteen years."

This report was followed by a modest research program administered by the National Science Foundation for $13.2 million in solar research. In the fall of 1973, Alfred J. Eggers, Jr., of the NSF, was made chairman of a special group that undertook a far more extensive report to the Atomic Energy Commission's Chair-person, Dixie Lee Ray, on the prospects for solar energy. Ray had been asked by President Nixon for priorities for energy research.

In summary, this report says:

> At an average energy conversion efficiency of five per-

cent, less than four percent of the US continental land mass could supply 100 percent of the nation's current energy needs. Thus, solar energy could contribute significantly to the national goal of permanent energy self-sufficiency while minimizing environmental degradation. In addition, this technology will be an exportable item for use by other energy deficient areas of the world. Although the full impact of solar energy probably won't occur until the turn of the century, the economic viability of several applications, e.g. heating and cooling of buildings, wind electric power, and bioconversion of fuels could be developed and demonstrated in the next five years. Ultimately practical solar energy systems could easily contribute 15 to 30 percent of the nation's energy requirements.

In most cases, photovoltaics being the primary exception, the development of practical systems will not require high technology. Thus, the research and development costs for solar energy should be very small in relation to the value of energy saved. Current estimates indicate that the value of the fossil fuel to be saved in one subprogram alone, heating and cooling of buildings, would equal the cost of the entire accelerated $1 billion R&D program seven years after practical systems become commercially available.

The 200-page report then sets forth specific plans for developing solar energy, spending $400 million in a minimum viable program and $1 billion in an orderly accelerated program over five years. The Eggers panel proposes to build solar power generating plants on land; rotor systems for wind power; plants to create methane from sewage; electric power plants at sea, and photovoltaic systems that can turn the sun's rays into electricity, either in the form of a large-scale power plant or in individual buildings. In addition the panel proposes additional development for mass production of systems for heating and cooling buildings.

But in making her proposals to the President, Dixie Lee Ray essentially rejected the study's conclusions, and instead of the minimal $400 million figure, proposed to spend only

$200 million in five years. The OMB subsequently raised the amount to $350 million, still below the minimal level.

Despite the new interest in solar energy, it is still slow to develop. The government's building program is small and relatively unambitious. Congress holds hearings and makes studies; but members of Congress, like most other government officials, generally take the position that solar energy can be used only as a back-up system for fossil fuels. This, of course, is the position of the major petroleum companies, whose general, although unstated, policy seems to be to exhaust one fossil fuel after another.

In 1973 the *Washington Post* carried extensive reports on the development of solar energy, then editorially recommended that Congress support an expanded program for its development. Two Congressmen, James Symington, chairman of a science and applications subcommittee, and Mike McCormack, chairman of a subcommittee on energy, wrote a letter to the *Post*, essentially arguing against the idea. Their position is typical: "We are enthusiastically supporting research and development programs in all areas of solar energy. However, we consider it our duty to emphasize that enthusiasm for solar energy should not inhibit in any way the more immediate and urgent programs in fossil fuel and nuclear research and development, upon which this country must inevitably depend for virtually all of its energy for the balance of this century."

Thus, while the government reports made it perfectly clear that solar energy offered an important contribution to the energy crisis, both Congress and the administration took initial steps to insure that its introduction would be delayed. They did so by relegating solar energy to more research and development; by establishing a pattern of research that would keep the scientists employed for years on expensive projects; and finally, by insuring that once solar energy was introduced, it would be through existing mass-energy systems and the companies that run these systems.

Although solar energy is an example of small-scale technology that can lead to breaking the hold of the big oil companies and utilities, the government slowly designed research programs that worked to insure that that would never happen. Oil companies and aerospace firms are major research recipients, and they work together with universities that receive funds from the NSF. Virtually the same as space program and Vietnam War research. The leading research teams are already well established, and include Harvard and Tyco Labs; the University of Minnesota and the Honeywell Corporation; Los Alamos Laboratory of the AEC, General Electric, the Inter-Technology Corporation, and Aircraft Armaments, Incorporated. In other solar studies GE is teamed with the University of Pennsylvania, Westinghouse with Colorado State and Carnegie-Mellon, TRW with Arizona State, and so on.

Most of this research is not concentrated on developing ways to institute solar systems speedily into heating and cooling systems for buildings, but rather concentrates on system aims, such as solar satellite systems and enormous electrical generating schemes based on "farms" of solar collectors in the desert—in short, systems that will mesh solar to existing inefficient, national electric-utility operations.

In 1974 the Congress had before it several different bills, which would have reinforced these arrangements with an expensive, self-perpetuating research and development program. The one bill that passed both chambers, originally sponsored by Congressman McCormack, would have NASA and the NSF spend $50 million over five years to build demonstration houses heated and cooled by solar energy, a token gesture benefiting upper-middle-class homeowners.

PART THREE
Resistance

12. The Georgia Power Project

The energy crisis of 1973-74 can be viewed as part of a long-term historical process that functions to reorganize the industrial base of the modern economy, in which industry and government work as partners.

Since the 1920s the terms of this partnership have been pretty much established by industry, with government serving as a passive agent. Recent attempts at legislative reform are part of this process, and only in a peripheral sense are they meant to change the basic structure of the industry.

Nonetheless, amidst this general industrial reorganization, there appeared in the early 1970s indications of serious, long-term resistance, especially at the local level. Much of this local resistance focused on electric utilities. Citizens' groups, unions, environmentalists, and other community groups have joined in efforts to gain control over their electric utilities, especially in California, Georgia, Vermont, Rhode Island, North Carolina, Pennsylvania, New York, Maine, Wisconsin, Arkansas, Florida, Indiana, Michigan, and Oregon. Three major efforts, in Georgia, Vermont, and California, are detailed in this section.

Perhaps the most determined of all these efforts is the Georgia Power Project, formed in 1972 by a group of young radicals in Atlanta. The project grew out of a strike against the Southern Company by construction trades unions. The Southern Company, a lage utility holding firm, is the major source of construction work in the Southeast, and it was employing nonunion labor.

Research into the company's operations was revealing

because it began to show how the holding company was draining funds from the different electric utilities it owned and how the entire operation, in turn, was bled by financial institutions in New York. All of this was accomplished by steadily increasing electric utility rates, especially the rates imposed on small residential users, who consume the least amount of electricity.

After the building trades strike, a group of about 20 people, some from the underground paper *Great Speckled Bird,* others working in the Southern Institute, formed the project with an eye to organizing citizens around the issue of utility rates.

In an early bulletin, the staff explained:

> The Georgia Power Project is composed of men and women who believe that the Georgia Power Company must be made to serve the interests of working people. We believe that electrical power is an absolute necessity in our society and, as such, access to it is a basic right.
>
> We find that the Georgia Power Company fails to provide adequate, economic service and, out of a desire for even greater profits, continually acts in ways contrary to our interests. This is seen in poor working conditions, anti-union activities, advertising to needlessly increase the use of electricity, pollution of our air, water and natural resources, discriminatory rate structures, racial and sexual discrimination, and in a host of other areas.
>
> We believe that the long-term solution to the problems posed by the Georgia Power Company is for the people most affected by the Company to assume direct control of its operations. We are socialists, working toward a time when the Company will be run and managed by its workers and consumers. We believe this is necessary not only for the Georgia Power Company, but also for many other economic and political institutions which so dominate our lives. This solution, however, is a distant one and one that many do not now accept. We say it here merely to make clear our beliefs and goals.

More immediately, our study and work have shown us that there are pressing problems facing us now demanding immediate action. Therefore, we have drawn up programs and alternatives to alleviate the most glaring abuses of the Company as they affect all our lives. We consider these to be minimal reforms and ones that many others who do not share all our beliefs and goals can support and help implement.

We understand that both our beliefs and those of people who disagree with us on some of our goals will change as we struggle to reorder the relationships of power in our society. We invite your participation in that process.

Although the project publishes a bulletin and is engaged in a variety of political activities, its major focus has been on the utilities commission and in court, where it struggles to establish new ground rules:

The Georgia Power Company for the second time in two years is demanding a huge rate increase to bolster its already gigantic profit margin (earnings are up 44% in the last seven months).

This increase must be approved by the Public Service Commission (PSC), the state agency empowered to regulate utilities. They are currently conducting hearings on this request.

The Georgia Power Project insists that this time the PSC exercise its responsibility towards the working people and consumers in Georgia and we charge that the Georgia Power Company has no legitimate standing before the Public Service Commission until it fulfills its obligations to all people in Georgia.

The record of irresponsibility of Georgia Power is clear. The Company has a long history of arbitrary cut-offs, poor service, and increasing rates, while catering to Georgia's corporate giants and the Northern Banks. It is the largest single polluter in the state, poisoning our air and water and thereby jeopardizing the health of all Georgians. It has a record of discrimination against Black people and women in hiring and promotion, and has recently made a blatant attact upon the state's working people by hiring a non-union construction contractor. The Company cries about an "energy crisis" and says it needs more money to meet projected increases in demand, while it turns around and spends our money for advertising to increase electrical usage.

The Georgia Power Company has no concrete programs to remedy these abuses. Therefore, the Georgia Power Project calls upon the Public Service Commission to protect the interests of Georgia's poor and working people by requiring the Company to fulfill the nine demands listed below before considering any further rate increases.

1. Electric power is necessary for the health, safety, comfort and convenience of all people. It is a basic human right. Present rate schedules of the Georgia Power Company, however, doubly penalize residential consumers: (a) residential rates are higher than those of industrial users who gain profit from the electricity they use; and (b) since the rate goes down as the residential consumer uses more electricity, the rate schedule penalizes those who use electricity only for essentials while encouraging luxury consumption.

Therefore the Georgia Power Project demands that: Present rate schedules be reversed so that industrial users pay more and residential users pay less; further, all rates must be placed on a graduated scale so that the price per kilowatt-hour increases as the amount used increases.

2. Currently, to obtain electricity, the Company requires a large deposit plus a service charge. This places an undue burden on poor and middle-income people. Moreover, Georgia Power often cuts off electrical service to customers, particularly low-income consumers, in a very arbitrary and inhuman fashion. This cannot be allowed to continue.

Therefore, the Georgia Power Project demands that: No deposits or service charge be required for electrical service; that all deposits currently held be returned with interest; that all involuntary cut-offs of electrical service be illegal; and that any refund due a customer be paid in full upon demand.

3. The Georgia Power Company, like other large corporations, reflects the sexual and racial biases of its owners and perpetuates these biases through its hiring practices. Statisti-

cally, 99% of its officers and managers (the people trained to run the company) are white males, while 73% of its clerical workers are women, and 78% of its unskilled laborers are black men.

For this reason, the Georgia Power Project demands:
The age, sexual, and racial balance at all levels of the company and its contractors reflect the balance of the work force in the entire service area.

4. The Georgia Power Company is conducting a campaign against the wages of working people in Georgia. By breaking its long-time policy of hiring unionized contractors and, instead, signing with a non-union contractor for the Yellowdirt Creek Plant project, Georgia Power is attacking the unions which traditionally have set the wage standard for Georgia workers. If this is allowed to continue, the wages of all Georgia's working people will suffer.

The Georgia Power Project demands that:
Georgia power company hire only unionized contractors.

5. The current policy of the Company is to increase demand for and consumption of electrical energy. This is accomplished through the rate schedule, advertising, offering savings to all-electric buildings, promoting sale of inefficient and needless electrical appliances. This policy is wasteful, shortsighted, and contrary to the interests of Georgians.

Therefore, the Georgia Power Project demands that:
The Georgia Power Company implement a program to stabilize electrical consumption and systematically eliminate waste of our natural resources and pollution of our environment. This program must include: (a) advertising by the company must be abolished, (b) the company must get out of the appliance sales business, (c) special deals on "all electric" construction must be made illegal.

6. The Georgia Power Company's rates are set by a formula which multiplies the *rate of return* times the *rate base* to give the *net operating income*. The larger the power company can make the rate base the more money they can take in at the consumer's expense. The rate base is supposed to consist of all the money that Georgia Power has invested in those facilities

used in the production and distribution of electricity. Unfortunately, much of the investment used in calculating the rate base in the last rate increase and in the current one has nothing to do with electrical power. Such items as replacement facilities for old, worn out plants are included alongside the plant being replaced. This means that the same generating capacity is being counted twice.

Another padding in the rate base is for antipollution equipment which produces no electricity and is the result of Georgia Power's choice of fuel and method of production, as well as their failure for many years to do anything substantial about its pollution.

None of these items have anything to do with producing and distributing electricity yet they are included in the rate base for determining the cost of power to the consumer. This unfair accounting must be halted.

Therefore, the Georgia Power Project demands that: Georgia power limit the rate base to the actual costs of production and distribution of electricity.

7. Georgia Power is presently beginning a massive program of expansion and new construction. That program will be financed almost entirely by the sale of securities. Those millions of dollars worth of securities, plus the substantial interest on them, will be paid for by the consumer, not by Georgia Power or the Southern Company (parent company of Georgia Power) stockholders. Since Georgia Power is committing us to pay off its debts, the Georgia Power Project demands:
A full public review of all sources of finance, both present and future.

By this we mean a full public disclosure of all of Georgia Power's financial arrangements, and public hearings to rule on all proposed financial dealings, including those with the Southern Company.

8. Further, the manner in which Georgia Power uses the money which it receives through bond sales, power bills, etc., is of immediate concern. Georgia Power is spending our money when it builds a new plant or when it advertises on

television. This is done to make more money for its owners and managers, not to provide better services, nor to improve the conditions of its workers and residential consumers. For just one example, Georgia Power has in the past avoided installing pollution control devices simply because they produce no additional revenue. The owners of Georgia Power make a little extra money while we have to bear the costs of the poisoned air and fouled streams. This cannot continue.

Therefore, the Georgia Power Project demands:
A full public review of how Georgia power uses our money.

By this we mean that before spending a penny of what is in reality our money, Georgia Power would have to submit a full proposal for public examination and approval. Working people and consumers must participate in the planning of all new construction and expansion to ensure that our needs are met, that our total environment is protected, and that the development of our state is rationally carried out.

9. Since Georgia Power, like all corporations, is headed by a Board of Directors whose main interest is profit, the emphasis is on benefits to the corporation rather than on the needs of its customers and workers. This is shown by Georgia Power's poor record on pollution, continual rate increases, and discriminatory hiring practices. It is evident that this record will continue so long as Georgia Power is controlled by those whose only interest is in maximizing profits.

Therefore, the Georgia Power Project demands that:
Representatives of Georgia Power workers and consumers be included on the board of directors in such a way as to insure that our needs become predominant over the interests of the profit seekers.

Repeatedly the project sought to instruct the public about how the utility financial system worked; in this case, how the Southern Company, a holding company and the parent of the Georgia Power Company, bled the Georgia ratepayers. The March 1974 issue of *Power Politics*, the project's newsletter, explains:

As many of you know, the rate increase decision was based on "coverage"—the ratio of the company's profits to its bond interest. The company complained that its earnings were too low to issue more bonds, and the Public Service Commission gave them the earnings they requested. This new income showed up for the first time in January, when Georgia Power's profits rose 54.4 percent. This allowed the Southern Company, which owns Georgia Power, to reduce its projected investment in new construction by $20 million. This money is now being provided by Georgia ratepayers.

Second, Georgia Power won a change in the fuel adjustment clause, which automatically increases your rates when fuel costs increase. While higher costs used to be passed on over a full year, now you are paying the company back over three months. Thus, your bill got three fuel adjustments bumps in January: (1) old cost increases still being charged back, (2) one-third of this month's coal and oil costs at old rates, (3) the higher price of new coal and oil.

It has long been the Project's position that these increased profits are all going to the same place—the big New York financial institutions that control the flow of investment capital. We know that eight of these firms own at least 17 percent of the Southern Company; we know that oil companies are also in the "top tier" of similar capital-intensive industries which meet their financial needs on Wall Street. It is reasonable to assume that these financial institutions own a large proportion of every aspect of the electric business—fuel supplies, equipment manufacturers, power companies and so on.

We are just beginning to understand that, by owning all phases of production, the financial institutions can control *where* they take their profits. Is regulation keeping electric company dividends at a level unacceptably low for them? Then fix prices at the equipment manufacturing level (The Federal Trade Commission has uncovered the second such conspiracy in a decade.) or boost prices on fuel. After taking profits there, go back to the regulatory commission and say that profits are going up everywhere else, so the utility needs a rate increase to attract capital.

If this seems too conspiratorial, look for a minute at our local holding company, the Southern Company, and how it shifts profits around. Southern owns 100 percent of the stock of the Alabama, Georgia, Gulf, and Mississippi Power Companies. Alabama and Georgia each own 50 percent of the stock of the Southern Electric Generating Company (SEGCO). Southern also owns Southern Services, a so-called mutual service company. That the service company is the nerve center of the system can be seen by the tasks it performs: "general executive and advisory services, power pool operations, general engineering, design engineering, purchasing, accounting and statistical, finance and treasury, taxes, insurance and pensions, corporate, rates, budgeting, business promotion and public relations, employee relations, systems and procedures and other services with respect to business and operations." (From a Southern Company stock prospectus.)

The system is also held together by people. Alvin Vogtle is president of Southern, chairman of the board of Southern Services, a director of SEGCO, and vice president of the power companies. But because such interlocks look bad, they are kept for the most part at the assistant secretary/assistant treasurer/assistant comptroller level. Documents on file with the Securities and Exchange Commission show a gentleman named E. Ray Perry signing for at least four of the six affiliated companies. Payments flow back and forth between the affiliates for services, power and the like, and are subject only to minimal regulation. Thus a company making a regulated profit may show apparently innocent cost items which in reality are transfer of profits to another regulated affiliate which is allowed a higher profit, or even to an unregulated affiliate.

SEGCO, which operates a mine-mouth plant generating power for Georgia and Alabama, appears to be a tool for removing Georgia-paid profits from Georgia regulation. While Georgia Power's payments to SEGCO for power are included by the Public Service Commission when determining rates, neither Georgia Power's dividends from nor capital investment in SEGCO is included in computing Georgia Power's rate of return. Thus Georgia Power sends on to Southern its own profits, at a regulated return, plus SEGCO's profits, at an unregulated return. Taking it a step further, SEGCO uses some of its profits to retire its stock rather than to pay dividends, probably to avoid taxes.

Southern is now proposing that Alabama build a new plant at SEGCO's site; that facilities at the site be divided between Alabama and SEGCO; and that Alabama operate the site as SEGCO's agent. Thus, in order for Georgians to determine how much their SEGCO power really costs, they must (1) get from Georgia Power's accounts to SEGCO's accounts, and (2) get from SEGCO's accounts to Alabama Power's accounts. Clearly, this is a wonderful opportunity to rip off Georgia ratepayers without having to answer for it. Another interesting feature of the agreement is that blue-collar SEGCO employees will become Alabama Power employees, but white-collar workers will stay with SEGCO. With Alabama doing the work and Southern Services doing the management, what is left for these folks to do but collect their checks—at the expense of Georgia ratepayers? Remember that this agreement is between two companies owned by the same people, one of which owns half the other (the application for SEC approval was signed for both sides by the ubiquitous E. Ray Perry). Now ask yourself—who is going to benefit?

All of this rigamarole is supposedly regulated by the SEC under the Public Utility Holding Company Act of 1935. Well, we'll see. The Georgia Power Project was denied a hearing by the SEC on one complaint, and we will appeal that. We'll ask for more hearings in the future. Hopefully we'll stop some of this manipulation by the Southern Company.

But remember, the Southern Company was only an example. It itself, and most of the rest of America's industry, is owned by the New York financial institutions, whose activities are almost entirely unregulated. And if Chase Manhattan Chairman David Rockefeller becomes Secretary of the Treasury, as predicted, who is going to make hay?

132 Resistance

We started out by asking why the electric rate increase was higher than expected. We ended up by asking who controls America's economy for whose benefit. Electricity and democracy, think about it—there's a connection.

In the 1972 rate case, the project objected to the company's rate structure on grounds that it discriminated against residential consumers, particularly those with low incomes, and that it promoted the use of electricity. The project requested hearings, and when they did not take place, went to the Superior Court of Fulton County. The state then said hearings would be held, whereupon the court dismissed the complaint.

But instead of holding hearings, the commission summarily approved the company's rate structure. The project went back to court, and the commission again agreed to hold hearings. Finally at hearings the project introduced detailed studies showing that the growth of energy consumption is lowest among small users and highest among above-average users of electricity. The project proposed an inverted residential rate structure and also sought to require marginal cost studies, rather than the fully allocated cost study submitted by the company. The commission's final order placed a higher percentage of the increases granted in the 1972 and 1973 rate cases on industrial customers; and within the residential class, required the increase to be graduated so that smaller users receive a smaller increase.

When the company applied to the commission for approval of a rebate plan under which contractors would be paid for installing electric heating and wiring sufficient to carry other high-load appliances, the project testified against the idea on grounds that it was promotional and would lead to strip mining. The application was denied.

During 1972 the project sought television time to counter the company's advertising. When two stations—one in Augusta, the other in Atlanta—refused, the project filed a fairness-doctrine complaint with the Federal Communications Commission. The FCC eventually found in favor of the project, and television time was granted.

When the company sought approval under the Holding Company Act for the issuance of $250 million of short-term debt, the project objected to it as excessive and establishing an unstable capital structure. The Securities and Exchange Commission granted hearings, the first such hearings at the request of consumers in over 30 years.

As a result of those hearings, the SEC staff argued in favor of limiting the Southern Company's dividend payout rate. If that happens, it will be an unusual and significant victory for the project, which all along has claimed that Atlanta ratepayers are bled by New York capitalists in the form of dividends and high interest costs.

The project has also testified before both state and city task forces on energy. It has lobbied Atlanta City Council members for a feasibility study of a municipal electric system, and it claims that about half the council members now are in favor of such a study.

In the spring of 1974, the project joined with the United Mine Workers to fight the Southern Company's efforts to import South African coal, a move that would cost U.S. miners' jobs. The Southern Company has contracted for 500,000 tons of coal from South Africa in 1974, and has orders for 920,000 tons in 1975 and over 1 million in 1976.

The project is an example of an organization that began by opposing electric utility rate increases, and then gradually moved toward planning and developing alternative programs for setting up electricity distribution. Writing in *Southern Exposure* magazine, Joseph Hughes points out:

> The Georgia Power Project is a primary example of a group which has tried to engage itself in the realpolitik of a region in an aggressive and positive manner. It has made use of the traditional regulatory agencies (the Public Service Commission, the courts, and the Federal Communications Commission), as well as the media, through press conferences, taped interviews and talk shows. The Project did not feel this was a compromise of its radical principles—but rather a way of proving itself to the people of the region, showing the intransigence of the institutions-that-be, and articulating the

potential of people gaining control over their lives and over capitalist institutions. As the system further deteriorates, time is ripe for developing more bold and imaginative strategies and tactics for attacking the concentrations of monopoly power. For the first time in 40 years, there is an opportunity for the Left in the United States to operate, not as a small isolated, sectarian group, but as a diverse and widespread political force. But in order to gain support and acceptance with people, the Left must show that it can win and that it will, in Andre Gorz's words, "restrict or dislocate the power of capital" over the lives of people.

Hughes then goes on to sketch out how a publicly owned utility might form the beginning of a wider socialist system:

The municipalities and rural electric cooperatives would form the backbone of the transitional, anti-capitalist system. Decision-making and control would be carried out at these primary levels by councils of consumers and electrical workers in each municipality or rural co-op. Models for setting up such local electric distributing systems are numerous in our own American history, as spoken of earlier. It is possible that the production and distribution of other services could then be developed with electrical service as a first step.

In order to coordinate these smaller backbone units, a larger regional authority would be developed. The regional authority would assure the reinvestment of profits inside of the region and also re-distribution of the decision making powers over those funds back to the backbone of the system—the municipalities and rural co-ops. With control over the regional electric systems, planning could be begun for developing a more well-rounded economic base and product mix to replace the current dependence of many parts of the South on one commodity or service, be it tourism, coal mining, timber, textiles, or agriculture. A program for "decongesting" the urban areas and breaking down the rural/urban contradictions could be started with the funds generated through electricity production.

As the municipalities gained in strength and financial staying power, they could start to levy increasingly progressive taxes on the private corporations which were operating in their area, and thus keep some of the wealth generated in the area within the region.

The project's campaign has been mainly legal, involving a series of interventions before the state utilities commission. As a result, the commission has denied the company its full requested rate increases. That in turn forced the company to curb expansion. That pleased consumers, whose rates are not as high as they might be, and environmentalists, who are anxious to stop utility growth.

While the Georgia Public Utilities Commission rejected the project's arguments against the proposed $48 million rate hike in 1972, the presence of the project encouraged the commission to grant an increase of only $18 million. In 1973 the project intervened to block a proposed $11 million temporary emergency rate increase. Again the commission rejected the project's basic contentions but did allow the increase on a kilowatt basis, as argued by the project, which resulted in an increase to residential customers substantially less than that to industrial customers. In July 1973 the company applied to the commission for an $86 million rate increase. The project intervened, and this time the public utilities commission more generally used the arguments advanced to restrict the company.

The project, for instance, had argued against inclusion of construction work in progress in the rate base and objected to the inclusion of additional revenues to support the company's overleveraged capital structure. The utilities commission declared that the company's capital structure was overleveraged and stated that it should not increase its payout ratio until the capital structure was improved. It ordered the company to revise its growth projections within six months and said the company should not advertise except to promote the conservation of electricity.

13. Lifeline Service

During 1974 both the Vermont and Massachusetts state legislatures debated proposals for a Lifeline Service, a plan to make basic amounts of electricity available at established low rates. The Lifeline idea was developed by Lee Webb at Goodard College in Plainfield, Vermont. Webb had been involved in various rate cases against utilities in Vermont, but he believed that most of the attacks on utilities were specialized and narrow. He wanted to come up with a scheme that would be easy for people to understand, a simple way of making public the fact that residential consumers are subsidizing big industrial users of power. In addition, he wanted to develop a proposal that could unite two important, but often divergent, political groups in the state: trade unions and low-income groups, which are interested in economic issues, and environmental groups, which want to reduce the demand for electricity. In August of 1973 Webb and Scott Skinner of the Vermont Public Interest Group developed the Lifeline concept. Webb writes:

> While the extent of electrical use may depend on income and style of life, virtually every home in Vermont requires, as an absolute minimum necessity, power for refrigeration and lighting. This basic need can not be reduced and there is no feasible alternative to the service. There is evidence to suggest that low users may be paying more than their share of costs of providing electricity and the debate on this subject has been heightened by a period of rapidly rising rates. Rates are rising for many reasons, but one prime cause is the escalating costs of building new power generating plants. And while a Vermonter using a constant amount of energy over the years has

not contributed to the need for this new power generation, that person is now being asked to pay for it in higher electric rates.

The problem is compounded by the fact that for increasing numbers of Vermonters, the rates are rising beyond their ability to pay. But since electricity is a necessity, the rate must be paid, usually by cutting back on other vital necessities such as food or housing.

A crucial beginning in the struggle to institute a fair and nonpromotional rate structure in Vermont is the establishment of Lifeline Service—an amount of electricity sufficient to meet basic needs, available at a price that citizens can afford.

We propose that residential customers of utilities in Vermont be charged a maximum of $10 for the first 400 kilowatt hours used monthly. This amount of electrical energy is probably sufficient for such necessities as lighting, refrigeration, ironing, and operation of furnace fans, etc.

Lifeline Service would apply equally to all customers of the utility. Each customer would receive the first 400 kilowatt hours at a rate of 2.5 cents per kwh. If the full 400 kilowatt hours were used, the charge would be $10. If a smaller amount were used, the charge would be proportionately less. Present minimum charges which vary from utility to utility would be eliminated.

The 400 kwh provided for in Lifeline Service would not be subject to rate increases or any purchase power and fuel adjustment clauses. The Service would also be limited to permanent Vermont residents, and would exclude vacation homes.

Lifeline Service is a feasible financial alternative to the present array of rate schedules which vary from utility to utility. It would inevitably force readjustment in the burden of rate-paying which at present falls heavily on the low user. If the costs of Lifeline Service were distributed equally by kilowatt hour on all users (industrial, commercial and residential) using 400 kwh per month, the burden would be shifted to large users and would not affect residential customers using up to approximately 1250 kwh a month.

Lifeline Service is not an attempt to legislate rate structures for all electric utilities, but rather provides a fiscally sound, equitable base to which future rate reforms will hopefully be added.

Lifeline was endorsed by several different groups including Vermont Alliance, a citizens' political organization; Low Income Advocacy Council, a collection of poverty-funded CAP agencies; a senior citizens' legislative committee; several groups of handicapped persons; vocational groups; and the Liberty Union Party, the third party in Vermont. Lifeline was then presented to the state legislature, debated, and reported adversely from the House Commerce Committee. (The vote was 6 to 5 against.). In the house of the state legislature, Lifeline was defeated 2 to 1. Between January and March 1974, Lifeline was a major political issue in the senate.

In Massachusetts the Public Interest Research Group introduced a Lifeline bill to provide immediate short-term relief to small and moderate users of electricity. Martin Puterman of PIRG pointed out:

> Lifeline would achieve this result by setting a monthly rate of no more than 3¢ per kilowatt-hour for the first three hundred kilowatts of electricity consumed by each permanent residential utility customer in the Commonwealth. The 300 kwh figure, though less than the average Massachusetts consumption rate of 520 kwh per month, could provide a minimal amount of electricity adequate for essential appliances, lighting, etc.
>
> Based on figures from the rate departments of the three largest utilities (Massachusetts Electric Co., Boston Edison Co. and Western Massachusetts Electric Co.) over 50% of current residential bills are for less than 300 kwh a month. On the basis of the companies' own figures, we estimated that 80% of all residential users in the Commonwealth will receive reduced bills under Massachusetts PIRG's Lifeline proposal. It should be noted that this rate would not be subject to any further rate increases or collateral charges (services, minimum

or other periodic charges). However, the 300 kwh supplied under Lifeline would not be exempt from fuel adjustment charges.

Thus, under the Lifeline rate structure, permanent residents of the Commonwealth who consume 300 kwh per month would pay at most $9 for this service, no matter which company supplies the electricity. Presently in Massachusetts, the cost of 300 kwh ranges from a low of $8.60 to a high of over $12.00, depending upon the company providing service. These figures are basic rates and do not include the cost introduced by addition of the fuel adjustment factor.

The feasibility of Lifeline, from the utilities' viewpoint, has been demonstrated. Our statistics document that the adaptation of Lifeline would not reduce the utilities' revenues. Moreover, the fact that five municipal companies presently charge less than the Lifeline rate for the first 300 kwh "block" of electricity underscores the practicality of Lifeline.

The savings which Lifeline can afford the average small user of electricity will not result in loss of revenue to the utilities. Rather, consumers of larger amounts of electricity will make up this additional cost by paying a marginally higher rate, estimated to be less than .2¢ per kwh, added equally to each kwh consumed after the first 300 kwh of residential use; and to all kwh of other metered service.

These minor increases are strongly justified by the following facts:

1) Under the present declining block structure, large users of electricity already receive a significant subsidy from smaller users, who pay a significantly higher per-kilowatt cost. This discrepancy varies for each system. For example, in the Boston Edison system, a user of 100 kwh per month pays an average of 5.6¢ per kwh; a user of 300 kwh pays an average of 3.9¢ per kwh; and a user of 1700 kwh pays an average of 1.9¢ per kwh. Thus, the average user in Zone 1 (Inner City Boston) pays a rate twice as high as the average owner of an all-electric home in the suburbs (not including the fuel adjustment charge).

Such declining block rate structures have come under

increasing criticism both from independent economists and government energy experts. David Freeman, former Assistant to the Chairman of the Federal Power Commission and former director of the Energy Policy Staff of President Nixon's Office of Science and Technology, has noted:

> Redesigning rates and requiring the larger volume users to absorb the coming rate increases would alleviate the hardship on small consumers and provide real incentives to achieve greater efficiencies. And after all, it is the volume users whose growth is a primary cause of the increased costs for which the increases are needed.

Thus, the minor alterations which Lifeline may have on the present declining rate structure are justified on equitable grounds and may be viewed as an added beneficial effect of the measure.

2) Numerous economic studies have indicated that the demand of large users tends to be far more elastic than that of small users. Large users faced with a marginal price increase can modify their normal patterns of use, employ alternative forms of energy, and thereby avoid the consequences of the marginal price increases. In marked contrast, the small user of electricity, with little control over his consumption, must absorb these increases.

3) Available data indicate that large users of electricity waste significant amounts of energy. For example, the Federal Energy Office estimates that industrial users can cut their consumption by 30 percent without affecting their operations. By increasing the rate per kilowatt hour, however insignificantly, Lifeline will provide an incentive to minimize the amount of electricity which is not put to optimal use.

4) Other states have already taken actions designed to achieve the results we seek under Lifeline. Notable among these measures are a series of actions which have shifted part of the burden of electrical rates from small to large users. For example, the New York Public Utility Commission required the Consolidated Edison Company to apply the bulk of its revenue increase to the last two blocks of electricity usage. And in

March, 1972, the Michigan Public Utility Commission imposed the entire amount of a granted rate increase on Detroit Edison's higher usage blocks.

Congressman Ron Dellums of California recently underscored the significance of rapidly escalating utility rates for low-income consumers, noting that Americans on fixed or marginal incomes are being forced to choose among necessities of life such as rent, food and fuel. In an age when nearly 100 percent of Americans depend on electricity as their basic energy source, we can no longer view electric power as a luxury; a minimum amount of electricity is a necessity of modern life.

The present energy situation has called attention to the need for action, which is clearly long overdue. Lifeline is a practical solution which will begin to meet this need by guaranteeing the availability of basic electricity at a price which most citizens can afford.

Lifeline was defeated in the Massachusetts legislature, but lawmakers in several states are proposing some of the Lifeline measures.

14. The Fight Against Pacific Gas & Electric

In California during the late 1960s, there was a growing campaign to break the hold of the Pacific Gas & Electric Company, which controlled electricity in the northern part of the state. The company's strength was due in large part to its control over the use of electricity from federal and state financed power projects. The story starts at the turn of the century, when San Francisco cast about for a secure water supply. The city looked into 15 different sites before Mayor Phelan filed for rights on the Toulumne River. But because this site was Yosemite National Park, and the proposed dam would flood the lovely Hetch Hetchy Valley, conservationists, including John Muir, objected.

With Congress unwilling to allow the damming of Yosemite, California Congressman John Raker proposed a compromise: to let San Francisco take water from Yosemite, but in the process generate and distribute low-cost hydroelectric power.

But the stipulation was that both water and power go directly to consumers and that no profits whatsoever from this unprecedented public grant go to private utilities.

As J. B. Nielands wrote in the *Bay Guardian*, a weekly San Francisco newspaper:

> The Raker Act was the Magna Carta for cheap public power. It was thought to be tightly drawn in the public interest and virtually impervious to subversion by private power trusts. Its basic intent was to establish a municipal power distribution system in San Francisco, but it also allowed the sale

of power to public agencies and recognized the prior claims on the nearby Turlock and Modesto Irrigation Districts.

However, the Act stipulated, in strict terms especially irritating to the private power lobby, that any attempt to transfer the water or power to a "person, corporation or association" for resale could result in revocation of the federal grant.

In developing water, San Francisco complied with the Raker Act, issuing bonds to provide the funds necessary to build an intricate system of pipes and tunnels that carry the water from the Sierras across the central valley, under San Francisco Bay into a reservoir outside San Francisco. The city acquired the operations of the Spring Valley Water Company for $41 million so that it could deliver water directly to its citizens through the company's distribution system.

Today the water department, successor to the private company, supplies Hetch Hetchy water to 1.5 million residents in the Bay Area. According to a recent grand jury report, "The operations of the Water Department are entirely self-supporting, paying from revenues the cost of all operations, maintenance, improvements, taxes, bond interest and redemption. . . .

"Water is supplied to city residents at rates cheaper than those in effect when San Francisco acquired Spring Valley Water Company more than 40 years ago," the grand jury report says.

Hetch Hetchy's first small hydroelectric generator, the Early Intake Powerhouse, went into operation in 1918. It was at once connected to the Sierra & San Francisco Power Company (later merged with PG&E). The Department of the Interior declared this accord illegal on June 8, 1923; but nothing was pressed, since only a small amount of power was involved.

The completion of the Moccasin Powerhouse in 1925 made a substantial block of hydroelectric power available. The city began to lay a steel tower transmission circuit in the direction of San Francisco. The lines were strung across the

144 Resistance

central valley, some 99 miles to Newark on the east shore of San Francisco Bay. There the lines stopped. PG&E had a substation at Newark; and in anticipation of new energy, the company had laid a high-voltage cable across the Bay to span the remaining 35 miles to San Francisco.

San Francisco had on hand enough cable to finish the transmission lines into the city. But suddenly and unexpectedly, the city announced it was out of money and could not finish the system. The two electric companies then operating in the city—Great Western and PG&E—refused to sell their systems to the city; and instead of using the power of eminent domain to acquire them, the city approved a contract on July 1, 1925, to hand over Hetch Hetchy power to PG&E at Newark.

At the time, the *San Francisco Examiner* said:

> It is a wrongful and shameful policy for a grant of water and power privilege in the Yosemite National Park area to be developed at the expenditure of $50 million by the taxpayers of San Francisco only to have its greatest financial and economic asset, the hydroelectric power, diverted to private corporation hands at the instant of completion; to the great benefit of said private corporation, and at an annual deficit to the city of San Francisco.

In the 1925 elections every supervisor who voted for the sellout was defeated. The new board determined to sell bonds to obtain sufficient funds to get its power back. The first bond issue for $2 million failed to receive a two-thirds vote, although it received a majority. PG&E beat down eight bond propositions between 1925 and 1941.

To defeat the bonds, PG&E spent $200,000 between 1932 and 1942. At Congressional hearings Harold Ickes said PG&E's strategy was to "spread throughout the city the word that the Raker Act could be easily amended" and to confuse the issue by saying the city "had been discriminated against" by the act.

As Secretary of the Interior, Ickes filed suit to throw out

The Fight Against P. G. & E. 145

PG&E's 1927 contract. The case went to the Supreme Court, which ruled on April 22, 1940, that San Francisco had been illegally disposing Hetch Hetchy power to PG&E for the past 15 years. It also held that the act required a "publicly owned and operated power system" in San Francisco.

In its decision the Supreme Court said:

> Mere words and ingenuity of contractual expression, whatever their effect between the parties, cannot by description make permissible a course of conduct forbidden by law. When we look behind the word description of the arrangements between the city and the power company to what was actually done, we see that the city has—contrary to the terms of Section 6—abdicated its control over the sale and ultimate distribution of Hetch Hetchy power.

In the 1950s an investigation by California Senator Clair Engle revealed that San Francisco was allowing irrigation districts to serve as conduits to transfer Hetch Hetchy power to Pacific Gas and Electric. Statistics from the Federal Power Commission showed that 24.7 percent of the power purchased by Modesto and Turlock "is currently and for a period from 1945 to 1953" was sold to PG&E. Forty-eight percent of this total was Hetch Hetchy power, the FPC said.

Nielands writes:

> As a result of PG&E's influence, Hetch Hetchy's formidable power output is dribbled away in a fragmented pattern that brings relatively little revenue to the city. Besides the irrigation districts, power is sold to several low-paying San Francisco industrial consumers, which are served by PG&E lines from its Newark and Warnerville substations. The city pays for transmission charges, including losses.
>
> City power is wheeled into San Francisco on PG&E toll lines and the company until recently levied an outrageous toll. PG&E buys Hetch Hetchy power at Newark for $2 million, then resells it to San Francisco consumers for $9 million, congressional testimony showed in 1941. Total overcharge:

$6,600,000. Multiply these totals year by year and you begin to get the dimensions of this steal from the city treasury.)

The Raker Act dispute was lost in the tumult of the Second World War. The last court order, in 1945, orders the city to comply with the terms of the act and to take back control over the electricity systems.

During the 1960s there was a renewed campaign to enforce the Raker Act, brought about in large part by the *Bay Guardian*. Its editor Bruce Brugman, and reporter Peter Petrakis researched the history, published repeated stories, attacked the company, and even went before the public utilities commission. Nothing happened.

Then, in 1973, a member of the San Francisco grand jury took an interest in the matter. Her name was Jean Sullivan, and Ms. Sullivan was no doctrinaire supporter of public power. Each year the grand jury surveys different aspects of the city's operations. She persuaded the grand jury to make a study of the Raker Act, and it subsequently issued a scathing report, stating the arrangement between PG&E and the city to be illegal and demanding implementation of the act.

After reading about the report in the newspapers, Richard Kaplan, a San Francisco trial attorney, filed a complaint in federal district court to compel both the city and the federal government to implement the act.

While the main argument of the grand jury report is legal, its force stems from underlying economic considerations. At the writing of the report in late 1973, PG&E had increased electricity prices three times based on increased costs of fuel used to generate electricity. The alleged fuel shortages threatened blackouts. Joseph Y. DeYoung, a vice president, warned, "If sufficient quantities of oil cannot be obtained in a timely manner, we could face brownout or blackout in 1974." If PG&E was going to demand one rate hike after another due to rising fuel costs, and threaten disruption of service to boot, then it made all the more sense for San Francisco to throw out the utility and at last lay claim to the cheap hydroelectric power Congress had specifically granted it in 1913.

In its recommendations the grand jury says: "We are not public power advocates. We do not propose taking over PG&E's electrical generating facilities or PG&E's gas distribution facilities. In fact, we advocate the continued purchase of power from PG&E to supplement our Hetch Hetchy power."

The report proceeds to lay out a program for regaining control of the electric utility under a lease system. During the period of lease the city could assess the true value of the equipment and determine what parts of the electrical utility plant it wants to own, and what part it will leave with PG&E.

The grand jury concludes:

> The acquisition by the City of its own electrical distribution system is the only way San Francisco can fully utilize for its own purposes the electrical power produced by its great Hetch Hetchy system. This is *NOT* a case of the City *acquiring* power rights. We have *had* them for two *generations*. It is time that the citizens should realize the full benefits of this enormous resource of energy which we *own*.

Across the Bay in Berkeley, citizens' groups staged repeated campaigns to oust PG&E and replace the company with a new sort of power system. As Joseph Petulla wrote in the *Nation*:

> Since government regulation has failed to check the colossal economic and political power of the private utilities, there is need for new models of local and regional control over our natural resources. We thought Berkeley was a good place to start a movement in this direction. For although there are already 2,000 public power cities in the country, few seem to be actively concerned about problems of radioactive waste transport and dumping from public plants; nor are they speaking to the question of the conspicuous consumption of energy. Most public power cities are content to rake in large revenue surpluses for their municipal treasuries, at the same time enjoying lower rates than neighboring districts serviced by private companies. . . .

Our group dreamed that city ownership could pave the way for local control of all these matters, perhaps even under cooperative ownership, so that everyone voted on policy decisions, established rate schedules, etc., or at least elected the board which determined policy.

But PG&E cleverly fought off this effort. In 1966 Walter Packard, the conservationist, sought to persuade the Berkeley City Council to authorize an independent study of the feasibility of municipal power. Instead, PG&E made its own study, costing $79,000, that showed the project would not be feasible. When the request was again made before the City Council in the summer of 1971, PG&E brought in an updated version of the study that reinforced its earlier conclusions. The council accepted the study.

At that point the citizens' group decided to draft an initiative directing the city to take necessary steps to operate its own electric distribution system. The first step would be to make a feasibility study. To put the initiative on the ballot, 7,600 signatures were needed. By February the group had gathered 12,000.

The City Council relented and adopted the initiative into law, ordering a feasibility study. Almost immediately a taxpayers' suit was filed to block the ordinance, but it was defeated. A Citizens' Referendum Committee appeared to begin a referendum petition to disqualify the initiative ordinance. Spokesmen admitted that the sole contributor to the committee was PG&E, which gave $1,000. PG&E then employed individuals to gather signatures. According to Petulla some were paid $2 an hour. Others were told their pay was conditional upon their obtaining six signatures an hour. The Referendum Committee succeeded, and the measure was decided in the April 1973 municipal election.

PG&E then fought the plan by forming a new committee called the "No on Eight" Committee. Its financial backers included the Southern Pacific Land Company, three California banks, and a dozen or so large local companies. Since a Pacific Gas and Electric voter survey suggested that the student vote would cancel out the conservative vote of the people living in the hills, it concentrated on the black community, where the balance of power lay. Petulla reports that a black worker said PG&E hired blacks at $62 a day to canvass black neighborhoods. There were radio commercials and bombardments in the black newspapers, all to the effect that if the measure passed, taxes and electric rates would go up, and money for child-care and social programs would decline as the system drained the tax funds.

Petulla argues that the vote was lost in large part because of the amount of money spent by PG&E: $250,000. Even so, 42 percent of the vote was for public ownership. The swing vote was a narrow 3,500.

In the spring of 1974 the public power groups re-formed for another try in Berkeley. Led by Edward Kirschner, a public power advocate, the group counted on 40 percent that would vote against PG&E no matter what. This time the undecided vote would be influenced by the energy crisis and rate increases of 50 percent.

The proposed system would be an ambitious experiment in public power. The new system would be run by an interim commission appointed by the City Council. The public system would have to demonstrate clearly an improvement over PG&E. Rates would have to be lower than existing rates. Smaller users in low-income brackets would have the cheapest rates, not the other way around, as is common. No property taxes could be used to support the system. It would be self-sufficient. Extremely strict conflict-of-interest regulations would be applied to the management and commissioners. They would not be permitted to have any relationship with PG&E or other utilities. They could not be employees, owners of securities, or directors of companies interlocked with the private utility.

As for the power source, a public-powered Berkeley would seek through joint ventures with other public power cities or by direct puchase to obtain geothermal energy (heat from the earth's subsurface). There's enough geothermal power in northern California to power the entire northern part of the state, and much of it is on federal or state lands.

By the middle of 1974, attacks against electric utilities had spread up and down the East Coast. In Philadelphia the Strike Committee on Philadelphia Electric Company (SCOPE) organized a coalition of community, labor, church, consumer, senior citizen, and environmental groups, with a combined membership of nearly half a million, to fight a 21-percent proposed rate increase. Citizens pledged not to pay their bills until SCOPE decided how and when. In Connecticut a coalition of labor unions and consumer groups persuaded 20 percent of the consumers of the United Illuminating Company to withholding the fuel-cost adjustment portion of their electric bills. A class-action suit in Ulster County, New York, against Central Hudson Gas & Electric sought to declare the fuel adjustment clause illegal. In Chapel Hill, North Carolina, the electric utility was turned over to a public organization, made up of elected officials of surrounding towns.

In Rhode Island a Peoples Public Utilities Coalition, an ad hoc organization of welfare rights groups and unions, got 6,000 names on a petition opposing a fuel escalation clause and so impressed the utilities commission that it employed the group to investigate utilities operating in Rhode Island. The group pressed for a referendum that would ask citizens to vote for the state to take over all electric utilities.

In North Carolina the United Mine Workers made a coalition with citizen's action groups that fought the Duke Power Company. The UMW was attempting to organize Duke's Brookside Mine in Harlan County, Kentucky; and to do so, the union sought to put the company under financial pressure wherever possible. To that end the union retained a firm of utility experts to funnel information to the North Carolina Public Interest Research Group and Carolina Action, two citizens' groups that have intervened in the state utilities commission to block Duke's proposed rate increases. The union also paid for ads in the North Carolina papers opposing the rate increase.

In Washington the Movement for Economic Justice coordinated activities of different groups fighting utility rate

The Fight Against P. G. & E. 151

hikes. The Environmental Action Foundation opened a utilities project to function as a clearing-house for groups that are fighting local utilities. The project's first major undertaking was the publication of a helpful handbook called *How to Challenge Your Local Electric Utility: A Citizen's Guide to the Power Industry*.

The booklet is a 112-page basic guide to the power industry for citizens and groups interested in learning more about their electric utilities and becoming involved in decisions affecting the power industry. It contains a detailed analysis of the power industry's impact on the environment and its control over regulatory agencies and politics. It is also a guide to citizen action, with "how-to" chapters on changing discriminatory rate structures, challenging promotional advertising, opposing new power plants and lines, promoting energy conservation, and challenging rate increases.

With publication of the booklet, the utility project shifted its emphasis to helping local utility fighters by providing them with information, research, and communication. Substantial aid was provided in fights in California, Vermont, and Georgia. The foundation also collected the most progressive decisions of state utility regulatory commissions to provide guidance for citizens' groups and other state commissions. If a local group faces a rate increase due to a fuel escalation clause, it can write the foundation, and the staff will send back the best brief arguing against the fuel escalation clause. On occasion the staff will testify as expert witnesses in utility cases and, more to the point, put local groups onto experts in these areas.

In a way, then, the Environmental Action Foundation project has begun to spread around the techniques developed in Georgia or in San Francisco, also keeping close watch on national political trends and legal activities.

While most of the reaction to the energy crisis was concentrated on electric utilities, the Northern Plains Resource Council, a Billings, Montana, group, waged a shrewd and impressive political operation to block strip mining in the state. Initially Northern Plains was made up of ranchers who were threatened by strip mining; but now it also includes environmentalists and professional people.

The council has won state legislation removing the right of eminent domain from coal companies, forced through a power-plant-siting amendment, and persuaded the Governor, Thomas Judge, to ask the legislature for a moratorium on water leasing. In 1973 it narrowly lost a bill in the state legislature's lower house that would have placed a moratorium on strip mining.

In Washington, Northern Plains works together with the Environmental Policy Center against strip mining and helped write the Mansfield amendment that passed the Senate in 1973.

The Northern Plains Resource Council helped launch other anti-strip-mine groups in Wyoming and North Dakota.

PART FOUR

Proposal for an Alternative Energy System

The groups described in the previous section represent an uncertain resistance to the elaborate plans for industrial reorganization that have been in progress for more than a decade.

They are admittedly piecemeal opposition. Taking over local electric utilities is not likely to mean much if, as in the case of the New England Petroleum Corporation, the international oil companies which provide the petroleum and shipping decide to change their terms. A narrow movement toward public power could function as an industrial public-relations program, resulting in inefficiencies and higher rates for citizens because of the public power authority's inability to drive bargains and make workable arrangements with fuel sources. Or more simply it could turn out to be another case of the public bailing out a sick industry. Control of fossil fuel is crucial to exercising any sort of control over existing energy systems. Control over planning is important in the introduction of alternative energy systems.

Another way to look at energy is to consider what ought to be, given the specifics of an area's political economy. For example, an inquiry we made into the Chesapeake region (the area comprising Maryland, Delaware, Virginia, the District of Columbia, and parts of West Virginia) made clear that any sort of coherent energy policy must take into consideration the political and economic framework set down in the seventeenth century, and covering much more than the narrow concern for energy as we now perceive it.

The Chesapeake region lies at the southern end of the urban industrial corridor extending from Boston to Richmond. It ought to be a bountiful area—rich in farm lands, interlaced by an intricate estuarine system of protected bays and rivers. But the region is only marginally self-sufficient. From the time of the earliest settlements in the seventeenth century, this region has served as a colonial resource bin, never developing its own economy. In the beginning it provided staples to England and the continent—first furs, then tobacco—later wheat and cotton; and during the nineteenth century, iron ore.

Efforts at self-sufficiency were deliberately thwarted. During the colonial period, laws repressed native manufacturing lest the populace grow independent of European imports. Intercoastal shipping might have tied the region to other parts of the East Coast, but laws prevented colonials from owning ships, lest trade be diverted from Europe. The fishing industry, always a potentially prosperous undertaking in the Chesapeake region, had its growth retarded by a provision of the Navigation laws which allowed only northern colonies to import salt directly from southern Euorpe. Without salt fish could not be preserved.

The region has a diverse natural transportation system of waterways, which in the eighteenth century were interconnected and extended through canal projects. But these waterways, which might have tied the area together in beneficial commerce, were used instead for hauling goods to ocean ports for export to Europe. Commerce within the region and between the region and others actually was discouraged by making interregional transportation rates higher than rates from the area to Europe.

In the nineteenth century the railroads laid down an intricate pattern of lines throughout the area. But the lines served to further ruin the area by hauling away resources to distant industrial centers and subjecting regional industry to national competition, which forced it out of business.

The financial panic of 1837 created additional interest in crop diversification, and the lower tidewater region increased its production of truck crops for export north. Many orchards were planted. But the trend toward diversification was checked by the coming of the railroads, whose main effect on pre-Civil War Virginia and Maryland was to broaden the plantation areas and to intensify staple agriculture.

Today the agriculture of the region is a caricature of a

functioning society. Grain grown in the region is consumed by chickens that are exported to New York and New England. Chickens to feed people in the area are imported from North Carolina. The region imports calves and raises them. Then the young animals are exported to the Middle West for finishing in feed lots, where they are slaughtered and reimported to markets in the Chesapeake area. The area grows peanuts which are exported for processing in Atlanta, and then reimported. Potatoes grown in Maryland are shipped to the Middle West. Meanwhile, the region imports the bulk of its potatoes from Maine, Idaho, and Oregon. Before the Second World War, the region was a major exporter of tomatoes. Now it imports tomatoes from California. In sum, this immensely rich agricultural area imports 75 percent of its foodstuffs.

In its broad outlines, then, the history of this luxuriant, rich region of the United States is marked by slowly encroaching poverty, unbelievable waste, ruinous agricultural practice, and blighted industry. The population sprawls along the throughways running from Baltimore to Richmond. Large segments of the population work at nonproductive jobs for the federal government. The major port at Norfolk, once the staging point for colonial export to the Continent and Great Britian, is now the staging point for colonial export to Japan.

Never in its history have the people of this region sought to lay claim to their own destiny. Not once in colonial or post-colonial periods did they succeed in developing an independent social existence. The Chesapeake region was always a colony, first of Europe, then of New York and Boston. The ships of the East India Company were exchanged for the railroads of Morgan and Rockefeller.

This history is especially instructive because it is often argued that once the fetters of monopoly capital are broken (through vigorous antitrust action, etc.) then Americans can be free to return to the country's early days of small business and rugged individualism. But even the most superficial economic history suggests the Chesapeake region never had a life of its own, that in one form or another, it was always the victim of outside merchants and financiers.

But things could be different. By changing the transportation system within the Chesapeake area, the uses of energy could be reduced and congestion lessened in the dense urban strip running between Baltimore and Richmond.

One quarter of all energy in the United States now goes for transportation. Along the East Coast, dependent on oil imports, half of those imports go for fuel in trucks, buses, and cars. One simple way to reduce energy consumed in transportation and create efficiency as well is to change from trucks to trains. On the basis of miles per ton, trucks use 4 to 6 times as much fuel as railroads and produce that much more air pollution. And while railroads can handle 2 to 3 times their present freight traffic without clogging up, most truck-bearing highways operate close to capacity. In the Northeast the congestion is likely to grow worse if the railroads carry out their plans to eliminate tracks, thus reducing freight hauling capacity. As Barry Commoner explains:

> Highways capable of carrying trucks take up about 51,000 square miles. About 28 percent of highway construction is needed to accommodate trucks; truck freight involves about 14,000 square miles of land use, or about 34 square miles per billion ton-miles of annual truck freight. In comparison, freight-carrying rail lines occupy about 3,800 square miles or about 4.7 square miles per billion ton-miles of annual freight. Similarly because cement-making uses a great deal of heat, the energy needed to construct a highway is large—about four times that needed for the same length of railroad track.

If the people of the Chesapeake region could be persuaded to reopen and modernize the railroads that interlaced the area in the late nineteenth and early twentieth centuries, then the necessity for autos, buses, and trucks could be reduced substantially. Railways were numerous throughout the region and on a map resemble a trolley system in a big city. They connect Baltimore, Washington, and Richmond and tie these metropolitan centers to the port at Norfolk. As a historic shipping point, Baltimore is well served with rail systems. The railroads also snake down through the rich farm lands of the Chesapeake bay area and pass through farm lands on their way back into the mountains to the mines. These old lines provide a basis for a future transport system. Where railroads previously were used to remove staples from the region, they can now be turned into a transport system that knits the area together, bringing food to urban centers from outlying agricultural sections, hauling in coal from the mountains, and so on. Surplus can be shipped out by rail lines to the North, South, or West.

The Chesapeake is blest with a natural transportation system in its waterways. The bay opens to the Atlantic. Its fingers reach inland to virtually every major existing center in the region. Before the coming of the railroads in the nineteenth century, residents tied together these fingers of water and extended them back into the hills in an intricate canal system. There is no reason why the canals, rivers, and bays cannot once again become a major means of transportation within the region. They are a natural, inexpensive way for hauling all sorts of goods and people. Together, canals and railroads, so widely used during the last century, offer a sound basis for a future transportation system.

A truly regional rail and waterway system could be an incentive for local self-sufficient agriculture. Over the long term self-sufficiency in agriculture is an important step in reducing energy consumption by reducing reliance on transportation and the middlemen processing industries, and of course, it can also play an important role in reorganizing society within the region. This won't be easy. As we have indicated above, the Chesapeake region in many respects is a backward, undeveloped area. Its resources never have been fully utilized to support a society within its own boundaries. But with increasing food processing costs, higher costs for transportation, and higher energy costs in general, there is likely to be fresh interest in finding ways to reduce the cost of food. Reorganizing agriculture is one approach.

A system of life-sustaining agriculture essentially means supplying the existing population with food from the Chesapeake. At its most basic level that means redirecting the flow of chickens from Boston to Washington and Baltimore, of potatoes from national markets to regional markets, and so on. It means establishing sectors of agricultural industry within the Chesapeake that no longer exist (for example, an enlarged cattle industry); and in general it means an intensive cultivation aimed toward producing a greater variety of crops at different seasons of the year. Perhaps most important, a regional, planned agriculture would result in preserving agricultural land from urban developers.

The schemes for transportration and agriculture sketched above will save a good deal of energy. About one quarter of all energy consumed in the United States goes for heating and cooling buildings. And much of that now depends on oil imported from Venezuela or the Middle East, natural gas which is in dwindling supply, or coal.

In the future much of this nation's heating and cooling energy can come from solar sources. The National Science Foundation conservatively estimates that half the total amounts of energy now consumed in heating and cooling buildings can come from the sun. If solar energy is combined with more efficient architectural design, energy consumption can be reduced by as much as 80 percent. Better insulation and siting of buildings to take advantage of the sun's rays; better ventilation; narrower, tree-shaded streets—all are proven ways of dramatically reducing the consumption of oil, natural gas, and electricity in the home. Cluster housing which takes into account solar energy as part of an entire energy system not only reduces consumption of fossil fuels, but also conserves land, which can be used for agricultural or recreational purposes.

But, you say, this is all fuzzy-minded utopian thinking without the slightest likelihood of becoming a reality. Maybe so. Certainly we have argued in this book that the political and economic forces which govern our energy systems are caught up in a sweep of history that seems likely to lead to ever more concentration, with the central state interjecting itself more often to defend the interests of entrenched industries.

Even so, as events of the last few years suggest, our energy system, indeed the entire economic system, is undergoing considerable strain and passing through a period of transformation. The shift of the large oil companies into fuels other than petroleum is one indication of that change. The rise of OPEC and the altered relationships with the big companies is another example. Pressures on American interests abroad may well lead to increased concentration, to a tightening of the ties between state and industry. But they may also result in different sorts of change. As we have illustrated in the previous chapter, there are political organizations struggling against the trends toward concentration. The Georgia Power Project, citizen campaigns against utilities in the West, the fierce struggle by ranchers against the coal operators on the northern prairies, are not isolated instances of environmental resistance. These attacks on business, which in the mid-1960s appeared in the narrow context of demand for consumer rights or liberal reform, now are beginning to take on serious ideological overtones.

Thus, within the Chesapeake region opportunities for serious systemic change present themselves day by day. Congestion in and around Washington already has led to the building of a major subway system. Partly because of the congestion, and admittedly for other political reasons, bureaus of the federal government are shifting their headquarters away from the central Washington city out into the surrounding countryside, leading to a population shift. Because of the increasing cost of energy, especially oil, it would seem realistic and shrewd for politicians in this urban strip to campaign for the development of new rail systems that can connect with the high-speed mainline service running from Washington to New York, on the one hand, and with Washington's ambitious subway scheme on the other.

Throughout this region, as in other parts of the country, environmental groups have fought against building nuclear

Proposal for an Alternative Energy System 161

power plants, refineries, and liquefied gas shipping points. In 1974 the leaders of the fight against nuclear power shifted their attack against nuclear power to argue for alternative energy proposals, such as solar energy. It seems likely that within five years, if not before, solar energy and other energy conservation measures will become an accepted part of energy planning. In Davis, California, for instance, the city council actually drafted building codes to implement energy conservation principles and prepare for the introduction of solar energy into housing developments. Since people are faced with increasing energy costs for household uses, the Davis experiment seems likely to spread.

And while agriculture is more concentrated than ever with prices steadily mounting, there is already in the Chesapeake region the beginning of an alternative system. A system of retail food co-ops can spread to encompass trucking outfits, links with producer co-ops of small farmers, and so on.

So these are not such utopian ideas. The foundations for implementing them already exist, although piecemeal in actual design and operation. As the cost of energy grows, and availability is limited, these alternative schemes take firmer hold and gain in momentum.

Any energy system which might develop in the Chesapeake region would unavoidably be interconnected with other parts of the country and other parts of the world. Oil comes from Texas and Louisiana and Venezuela. Natural gas is transported from the Gulf of Mexico by pipeline and from Algeria by ship. Even under a system of self-sufficient agriculture, grain still would have to be imported from the Middle West. A surplus of fish would be traded abroad, and so on.

During 1974 a group at the Institute for Policy Studies sought to develop the outlines of an overall plan or system that would change the existing energy system throughout the country. In doing so, it obviously would affect other areas of the political economy. The resulting scheme calls for the creation of a new kind of energy system—a network of democratically constituted local, regional, and national energy organizations. This system would have the authority to produce, transmit, and distribute energy in the nation.

The system would be based on several principles, including the following:

1. The nation's natural resources belong to all the people. (As Leonard Rodberg pointed out, "With the limits of growth clearly in sight, with less than 40-years' supply of oil and gas left in the United States, these mineral resources have become precious national treasures. Just like our national forests and our rivers and streams, they must be subjected to rigorous public control, so their use can be regulated, their consumption curtailed, and the public's interest placed first in the making of energy policy.")

2. Each citizen shall be assured a fair share of the energy made available to the American people.

3. Whatever system is developed, it must be firmly rooted in local popular control. Thus, regional and district agencies, created under the plan, should be involved in every stage of the preparation of the national energy plan.

4. All information regarding the activities of every energy agency, all reserve statistics, and data on energy consumption should be publicly available on a timely basis, to facilitate the fullest possible participation by the public in the preparation of the plan, and in its subsequent implementation.

5. The prices of energy products should be set as low as possible consistent with the costs of production and the ecologically sound use of the nation's resources, including not only energy resources but also air, water, land, and other natural resources.

6. There should be the minimum possible consumption of nonrenewable resources. Where possible, the energy plan should be coordinated with other national planning aimed at reducing the consumption of nonrenewable resources.

The heart of the plan involves creation of a new local govern-

Proposal for an Alternative Energy System

mental unit to establish and administer energy policy—the Public Energy District (PED). This would be a new sort of municipal corporation, a political subdivision within a state. The idea is taken in part from Lee Webb's work on a model energy scheme for Vermont. In part it is based on historical experience in the state of Washington. As David Whisnant recently described the Washington experience in the magazine *Peoples Appalachia:*

> In concept the public utility district is relatively simple. Normally a PUD law authorizes a publicly controlled body to issue revenue-producing bonds, receive and disburse funds, acquire real estate (by condemnation if necessary), construct dams and other power generation and distribution facilities, and sell electric power. Many PUDs in the northwest are distribution facilities only, buying their power from the Bonneville Power Administration. All PUDs pay a specified portion of their receipts into the general revenue funds of their counties. As non-profit enterprises, they are able to supply electricity to their customers at about half the rate charged by private utilities, while paying off their own indebtedness to bondholders.
>
> The public utility district mechanism quickly proved capable of achieving spectacular results in poor Washington counties. Tiny Lewis County, with a population of 35,000 farmers, loggers, and cattlemen and no industry, now operates a $2 million-a-year PUD which provides nearly a quarter of a million dollars a year in revenues for the county—including $125,000 per year to support its public schools. Chelan County, also quite small, started its PUD in 1936 and purchased its first transmission lines nine years later. Within the next few years it bought out some existing power systems, built a 249,000 KW generating facility at Rock Island, and financed construction of its own Rocky Reach dam by selling $263 million worth of revenue bonds. The 800,000 KW Rocky Reach project is a model of activity in a public interest; its powerhouse even includes a museum of artifacts excavated during construction of the dam. Power from Rocky Reach,

available by 1961, attracted manufacturing installations by Alcoa, Dow Chemical, the Vanadium Corporation, and others. By 1967, 22 Washington PUDs were supplying electric power to 280,000 customers.

Under the plan, voters within a proposed public energy district would request a referendum on the establishment of such a district in a general election.

Directors of the PED would be elected at the polls as part of regularly scheduled elections, with standards set for local geographic and worker representation.

A public energy district would have the power of eminent domain, but not the power to tax.

The public energy district is the basic unit within the proposed system of local, regional, and federal energy planning and administrative bodies. It would conduct planning; carry out research and development, produce oil, gas, coal, uranium, etc.; design and manufacture solar collectors; build oil refineries; lay pipelines; and operate and construct electric generation systems—all of the functions now carried on by the different energy industries or fragmented public or nonprofit bodies.

It is anticipated that the district would continuously debate energy policy and establish and administer policy for the region. It would set utility rates and priorities for the end use of fuels.

The district is meant to be a powerful political and economic organization. For example, if an automobile manufacturer sought to open a plant within a public energy district, it would first have to submit to a detailed plan of operations to the PED, whereupon the directors would initiate hearings on the advisability of building such a plant, initially taking into account the plan's impact on energy and the environment. But as the PED developed, it might also go further, inquiring into the energy efficiency and usefulness of the end product, i.e., car, truck, or bus; the effect of the plant on employment and transportation within the PED; its environmental impacts, its effect on economic growth policies;

and in other ways looking into the beneficial and adverse effects of constructing the factory.

Within the different operations of the district, workers would manage and operate the facilities, although the overall policies would be determined by the district board or council, which of course also would include workers.

Regional Energy Boards

Each public energy district would send a representative of its board to a regional energy board. The federal government has developed 10 multistate regions for the purpose of administering its different programs; and while these regions are arbitrary, the plan uses them as a basis, at least tentatively.

(There are several different sorts of federal regions, including 6 large "depressed areas" defined by the Economic Development Administration; 25 metropolitan administrative areas, called Federal Executive Boards; and 10 overall administrative regions that cover the entire nation and its territories. Under Nixon the major emphasis was to develop the 10 regions, and the Departments of Labor, HUD, and HEW were all committed to similar regional concepts and often had offices in the same building in the same city. The cities that served as sort of regional capitals were: Boston, New York, Philadelphia, Atlanta, Dallas, Denver, Chicago, San Francisco, Seattle, and Kansas City. Nixon set up a regional council where representatives of each agency involved has a seat.)

While the public energy district would administer energy resources on a day-to-day basis, the regional board would allocate resources within the total area.

The Tennessee Valley Authority provides an idea of what a regional organization might be like. Since its origins in 1933, TVA has sought to mesh together different aspects of resource planning, electric power, agriculture, industry, fertilizer production, navigation, flood control, recreation, and conservation. It conceived of the immediate job as not merely to build dams and reservoirs, but to put people to work. It did not contract for the workers but hired them directly, building them communities and attending to their health needs. It was an important force in reinforcing existing state and local governments by delegating tasks to these governments on a contractual basis. Its free technical services helped raise the level of state and local services.

Even though it was entirely surrounded by hostile corporations and a federal government that reinforced those corporations, TVA became an immensely important economic force, far more so than is often recognized. It should be remembered that TVA's electrical production program literally made possible the nuclear industry. Without the vast quantities of electricity produced by the combined coal and hydroelectric plants of the Valley Authority, the Atomic Energy Commission's uranium enrichment plants could never have operated. In providing that electricity, TVA literally reorganized the coal industry. It introduced the concept of long-term contracts, was an important factor in mechanizing the coal industry, and became the single largest purchaser of coal, a vital factor in the market. It also introduced a modicum of sanity into the electrical utility industry through its interlinks with other private systems in the South and the southeastern mountains, particularly the American Electric Power Company's operations. Despite the vitriolic attacks made upon TVA by private power, the Valley Authority, through these entities, made the private systems stronger and more stable.

The tragedy of TVA is that because it became so much an instrument of national economic policy, it has been placed in a position of turning against its own constituency on the strip-mine issue. Because of its policy of providing low-priced electricity, the Authority seeks out coal at the lowest prices, and hence trades heavily in strip-mined coal from Appalachia. Strip mining is ruinous to the entire region and in effect, by buying the stripped coal, TVA turns its own constituency against it.

A similar situation developed around nuclear power.

TVA reorganized the coal industry to provide the electricity to enrich the uranium necessary for hydrogen bombs and nuclear power plants. In doing so, it was answering the dictates of the national military, which was anxious to perpetuate nuclear technology.

Under this new proposal the possibility that national policy would dominate would be greatly lessened by grounding the policies of a TVA-like authority in the local districts, which in this instance would include the strip-mined areas, and it could not become an instrument of top-down federal policy.

National Energy Organization

The purpose of this board or agency would be to coordinate the ideas and plans of the different regions. It would be an important organization, providing the point of contact with the federal governmental apparatus and the Congress.

It would have several principal functions. Perhaps its most important function would be to act as trustee of the nation's natural resources, allocating scarce resources to regions for distribution, according to plan, to localities.

In principle, all natural resources of the nation ought to be public, and not given solely to any corporation for exploitation on its own terms. But, as with all other aspects of this plan, there need to be transitional steps. Here is one good example.

The national agency could take over from the Interior Department administration of those territories already in the public domain, that is, areas specifically removed from commerce by the Congress for the purpose of the general public good. These federal resources include an extensive amount of mineral fuels. The estimates vary. According to a common estimate, over 50 percent of the fossil-fuel energy resources of the United States are in the public domain territories. Some estimates place the amounts as high as 80 percent. According to the Ford Foundation's Energy Policy Project report, about one-third of the remaining domestic oil and gas resources are estimated as likely to be found in the Outer Continental Shelf, which is part of the public domain. In 1972 the Outer Continental Shelf lands produced 10 percent of the domestic oil and 16 percent of the domestic gas. Estimates are inadequate and need to be fully done. Oil shale is almost entirely controlled by the federal government. About one-half of the domestic coal in the West is under federal control. About 85 percent of the strippable low-sulphur deposits are in the public domain. About half of the nation's geothermal resources are on public land. An estimated 50 percent of the domestic uranium supply is in the public domain.

And of course these estimates do not include the huge areas of Alaska that have already been leased by the federal government to oil companies, or the state-controlled lands.

Under one concept a transitional scheme would be to place these important resources, already in the federal public domain (and in one sense "nationalized") within the control of the national agency, whose regional constituents could then make initial plans and coordinate national policy based on this resource base.

Eventually the basic idea would be to widen the concept of public lands so that all natural resources, including mineral fuel resources, would be considered public.

In principle, then, all energy sources would come under the public scheme.

In addition, the national organization should have a planning staff that functioned as a public research and development center serving the different regions. Probably this staff would conduct the mapping and resource estimates that are now carried out by private industry.

The national organization would take over the functions of the Federal Power Commission and the other regulatory agencies. For instance, it would establish all interstate rates and end-use priorities for energy, and arrange for international trade.

As the history of the modern energy industry instructs, again and again large corporate interests—the Standard Oil trust, its successor companies, the Morgans, Insulls,

Rockefellers—controlled different sectors of the industry through control of the transmission facilities. Rockefeller initially built his monopoly through control over transportation. In the 1930s the Morgans and the Rockefellers controlled the natural gas business by dominating the pipelines. In California today the major companies control the industry by ownership of pipelines. The inefficient electrical systems that cause brownouts and blackouts are due in large part to the refusal of private companies to transmit public power and interlock their systems with public power systems. Ownership of tanker fleets, the largest navies in the world, are controlled by seven major oil companies. Railroads refuse to haul coal from one market to another, thereby contributing to shortages.

Transportation of energy is absolutely crucial to its ultimate control. Therefore, under the plan the major interstate transportation facilities should be placed under direct control of the national energy board. This is a crucial part of our long-range plan.

The plan would have the national board in a staged process over 10 years acquire outright control (obtain 51% of securities) of the major interstate natural gas and oil pipelines and electrical transmission systems.

During this 10-year period, the national energy board would lease and operate those portions of oil, gas, and electrical transmission systems necessary to transmit energy from public domain territories to the different public energy districts. The terms of the leases would be negotiated between the board and the companies.

The lease period would provide an effective test of the systems, and the energy board could determine which parts of the transportation lines could be used in its developing interregional system.

In the case of interstate commerce in energy that is transported by water, rail, truck, or airplane, the energy board would establish rates and prescribe general policy.

While the national board would determine policy and establish rates, the actual work would be carried out at the

local level by the PEDs. Neither the national energy board
nor the regional boards would maintain sizeable bureau-
cracies. All work, including planning, bookkeeping, hearings,
and investigations, would be conducted by the PED staff.

The national energy board would regulate commerce in
energy between regions. Commerce within a given region,
among the public energy districts, would be governed by the
regional board. Commerce within the public energy district
would be regulated by that board.

Planning

As the brief history of the oil and coal industries indicates, the crucial element in the industry's control of public
resources and of federal governmental policy is planning.
Systematically, since the early 1920s, the federal government has given over to industry access to natural resources
and has refused to plan these resources.

The central, most important step in breaking apart big
capital from the federal government is to remove planning
from the industry. The representative federal board as en-
visioned in this plan would conduct routine, careful mapping
of the nation's mineral energy resources, including geophys-
ical assessments, shallow- and deep-core drilling, environmental tests, aerial and space surveys, mapping and testing
of the nation's coal, etc.

As with other parts of the proposed system, the actual
work would be carried out within the different energy districts under contract from the federal and regional boards.

Federal money designated for planning would be ear-
marked for use first by local energy districts, and secondly
through contract with nonprofit groups within the localities.

Where the money was spent on private industry, it
would go to locally owned and managed small business.

Research and Development

History instructs that private industry cast off its obsolete appendages on the state. When Penn Central collapsed,

the banks and other creditors persuaded the federal government to take over the passenger service, while leaving the healthy, profitable part of the railroad to private enterprise. Lockheed received the same sort of special treatment.

Instead of involving itself with the most dynamic, socially beneficial aspects of industry and commerce, the government—state, federal, or local—acts as the passive agent for big capital.

This is now repeating itself in energy. Con Edison, that pathetic electric utility, after bilking New York consumers in 1973, blackjacked the state of New York for $500 million in loans under threat of going out of business and discontinuing service.

In solar energy the same ruinous pattern is occurring. Federal agencies—HUD, NASA, the NSF, the AEC—quarrel over who gets what in solar research. Whichever federal agency gains control then siphons the research money off to the parasites that surround it—big universities and big business. The funds are used to reinforce the very system that has produced the energy crisis.

The plan is to break this pattern in several different ways. As the government and industry experts freely admit, solar energy is a regional technology. It offers different prospects for different parts of the country: hot sun in the deserts; heavy, continuous winds on the coasts; and so on. Natural phenomena offer energy prospects to be carefully used.

Under the plan the national energy board, which would be truly representative of the regions, would make research and development policy. The actual work of effecting the policies by the board would be carried out within the public energy districts. Funds spent for the development of solar energy would first go to the district, and after that to small business within the region.

One scheme for a local, nonprofit solar company might look like this:

Each not-for-profit company would be owned by everyone living in its sales area—that is, by the potential customers of its locally sold products. Seed money to start these operations would come from local taxes—tax monies

172 Proposal for an Alternative Energy System

collected in the sales area. Any federal monies supplementing or matching local funds would be seen as a portion of the locality's federal taxes and would exceed neither the amount of local funds nor $10 per person in the sales area. All funds would be no-interest loans.

Persons wishing to form a community-owned solar facility would present to everyone living in the "sales area" (neighborhood, community, municipality, county, state, region) a proposal stating:

   1. The projected costs and sales income over the first five years (including number of employees, wages, price of product, loan payment schedule).

   2. The amount of local tax monies needed and the amount of federal funds sought.

   3. The projected number of sales of individual products per year for five years.

   4. The geographic limits of and the number of people in the sales area.

   5. The types and quantities of products to be made per year for five years.

   6. The ecological impact on area citizens.

On the basis of this proposal, a referendum vote would be taken in the sales area to determine whether or not the facility would be established. The proposal and product designs would then be submitted to the National Bureau of Standards to determine that the proposal conformed to NBS-established guidelines of price, production costs, and product design.

The company would be owned in common by all individuals living in the sales area, and major decisions on company operations would be made on a one-person-one-vote basis at an annual owners' meeting. A board of trustees would be selected at this meeting. One-third of its members would be elected by the workers in the facility, one-third elected by the community, and one-third selected from the community at random (jury style). These people would meet monthly, would receive jury-duty fees for their once-a-month services, and could serve no more than three years.

The day-to-day working of the facility would be determined by the workers in the facility, within the production and price guidelines of the original proposal. Such facilities would have no "managers"; workers would determine issues of working conditions, health and safety, speed of production, hiring and firing, etc. When problems arose that workers could not solve within the workplace, or that might conflict with established production-price guidelines (i.e., ecological problems, complaints from surrounding residents, rise in cost of raw materials, revised budgeting for additional employees to meet higher production demands), the problem should be placed before the board of trustees at its monthly meeting. If major policy changes were called for—or if members of the board could not agree on policy changes—any member of the board might call a special meeting of community-owners to decide the question by a general vote.

This model of a community- and worker-owned solar production facility would be instituted immediately to produce most solar devices. In areas where additional solar research may be needed (solar-powered automobiles, municipal services, etc.), research funds should be available only to nonprofit groups, community-owned research organizations, or individual inventors, and not to private, profitmaking corporations. All designs or solar models resulting from such federally funded research would be the property of the American people, and construction designs would be openly available to any individual wishing to build a device for personal use and to any nonprofit, community-owned facility; but the devices could not be produced commercially by privately owned or profit-making companies—thus insuring that no profit or research costs would be attached to products that already partially "belong" to the American people as a result of their research and development financing.

Private Enterprise

Under the proposed scheme private enterprise would have a tightly circumscribed role. The emphasis would be on public agencies and to a lesser extent on small business.

174 Proposal for an Alternative Energy System

Under the principle that all natural resources are in effect within the public domain, the operations of the major energy industries—coal, oil, natural gas, uranium—would be brought under the control of the national and regional boards, whose members would determine how they were to operate and at what rates.

As indicated above, the transportation systems would be controlled (i.e., owned at least 51 percent) by the energy board, which would place a majority of its members on the board of directors.

While theoretically the operations of most of the big oil companies are now subject to control by the federal government because they operate within the public domain, in reality this is not the case. Since the 1920s the government has been intertwined with the companies, acting as a passive agent.

But an energy system as outlined above, grounded in a representative system and in control of the transportation linkages, ought to be a strong force in redirecting energy planning.

Chapter Notes

PART ONE: The Roots of the Crisis:
A Brief History of the Energy Industry

Chapter 1. The Standard Trust

See Ida M. Tarbell, *The History of the Standard Oil Company*, briefer version, ed. David M. Chalmers (New York: W. W. Norton, Norton Library ed., 1969). For a brief and excellent summary of Standard Oil's growth *vis-à-vis* Wall Street, see John Moody, *The Masters of Capital*, published as Part I of *Great Leaders in Business and Politics* (New Haven: Yale University Press, 1919).

Chapter 2. The Modern Industry

See Temporary National Economic Committee, 76th Congress, 3rd session, "Survey of Shareholdings in 1,710 Corporations with Securities Listed on a National Securities Exchange," Monograph No. 39 (Washington, D.C.: Government Printing Office, 1941); also Gerald D. Nash, *United States Oil Policy* (Pittsburgh, Pa.: University of Pittsburgh Press, 1968), Chapters 4–6.

Chapter 3. The International Petroleum Cartel

See 82nd Congress, 2nd session, Senate, Select Committee on Small Business, "The International Petroleum Cartel," a Staff Report to the Federal Trade Commission, August 22, 1952. Also George W. Stocking, *Middle East Oil* (Nashville, Tenn.: Vanderbilt University Press, 1970), Chapters 1 and 2.

Chapter 4. World War II

See Leonard Mosley, *Power Play* (New York: Random House, 1973), Chapters 11 and 12. See also Gerald D. Nash, *op. cit.*,

Chapters 8 and 9, as well as the FTC cartel report. The state Department memo is quoted from Joseph Stork, *The Middle East and the Energy Crisis* (New York: Monthly Review Press, 1974), Chapter 3.

Chapter 5. Oil and the Cold War

See the testimony by George McGhee, in 95th Congress, 2nd session, Senate, Committee on Foreign Relations, Hearings before the Subcommittee on Multinational Corporations on Multinational Petroleum Companies and Foreign Policy, January 28, 1974, p. 83. See also George W. Stocking, *op. cit.*, Chapter 6. Details of cartel prosecution and arguments between the State and Justice Departments are taken from 93rd Congress, 2nd session, Senate, Committee on Foreign Relations, Subcommittee on Multinational Corporations, "The International Petroleum Cartel, the Iranian Consortium and US National Security," February 21, 1974, pp. 1–158.

Chapter 6. Coal

Most of the historical material on the early coal industry is from Howard Eavanson, *The First Century and a Quarter of the American Coal Industry* (Baltimore: Waverly Press, 1942). Statistics are from several annual editions, between 1915 and 1950, of the *Keystone Coal Manual* (Keystone Coal Association, Pittsburgh, Pa.) and the *Minerals Yearbook* (Washington, D.C.: Department of the Interior); and from "Coal Industry Structure," in *Mineral Facts and Problems* (Washington, D.C.: U.S. Bureau of Mines, 1947). Information on the anthracite industry after World War II and the nationalization debate is primarily from C. L. Thompson, "Should the Coal Mines be Nationalized?" *Forum*, November 1949, pp. 285–95; and "Need for a National Coal Policy," *New Republic*, November 28, 1949, pp. 5–7.

Chapter 7. Natural Gas

See Temporary National Economic Committee, "Investigations of Concentration of Economic Power," Monograph No. 36; and Gerald D. Nash, *op. cit.*, Chapter 2.

Chapter 8. Electric Utilities

Information on early holding companies in the electric utility industry is from John Bauer, *The Electric Power Industry* (New

York: Harper and Bros., 1939), and Kenneth Field, "The Public Utility Holding Corporation" (thesis abstract, University of Illinois, 1928). Pre-1925 data on public ownership and J. P. Morgan is primarily from Harry Lee Williams, *The Power Trust Vs. Municipal Ownership* (Chicago: Public Ownership League of America, 1929), pp. 1–55.

Recent statistics on public ownership come from Vol. 27 of the annual *Directory of Public Power*, published by the American Public Power Association, Washington, D.C. Other statistics are from FPC reports. Edward Vennard's *Government in the Power Business* (New York: McGraw-Hill, 1968) provided some information on the operations of TVA and other federal power projects. The Temporary National Economic Committee monographs cited are "Investigations of Concentration of Economic Power," Nos. 32 and 36.

Information on the utility industry during the past 20 years is primarily from 83rd Congress, 2nd session, Senate, Subcommittee on Antitrust and Monopoly, "Monopoly in the Power Industry," Interim Report, 1955; 84th Congress, 2nd session, House, Committee on Government Operations, "Private Electric Utilities Organized Efforts to Influence the Secretary of the Interior"; and 93rd Congress, 2nd session, March–May 1974, Senate, Committee on Government Operations, "Disclosure of Corporate Ownership" (Washington, D.C.: Government Printing Office, 1974).

PART TWO: The Energy Crisis of 1973-74

Chapter 9. The New Industry

For a discussion of joint ventures and Mattei's influence, see Leonard Mosley, *op. cit.*, Chapters 19 and 20. The Joseph Stork quotes are from *The Middle East and the Energy Crisis*, Chapter 5. The best source of information on bidding and joint ventures is the testimony by Dr. John W. Wilson, 93rd Congress, 1st session, Senate, Judiciary Committee, Antitrust Subcommittee, June 27, 1973. Also see the Final Report of the Ford Foundation's Energy Policy Project, 1974.

For information on the California oil cartel, see the published hearings of the Subcommittee on Crude Oil Pricing of the Joint Committee on Public Domain of the California Legislature. Otto Miller's deposition was taken January 4 and 18, 1974, and filed at the Superior Court, Sacramento, in the matter of the petition of the

178 Chapter Notes

Subcommittee on Crude Oil Pricing of the Joint Committee on Public Domain of the California Legislature, No. 241,392. See also Kenneth Cory's testimony, 93rd Congress, 2nd session, House, Select Committee on Small Business, April 1974.

Information on the coal industry is taken from James Ridgeway, *The Last Play* (New York: E. P. Dutton, 1973), beginning on p. 27. See also "The New Energy Barons," by Matt Witt, in the *United Mine Workers Journal*, July 15–31, 1973. The best single source for a summary of events on the northern Great Plains is "The Agony of the Northern Plains," by Alvin M. Josephy, Jr., in *Audubon Magazine*, July 1973.

Chapter 10. Oil Shortages

See 93rd Congress, 1st session, Senate, Government Operations Committee, Permanent Subcommittee on Investigations, "Staff Study of the Oversight and Efficiency of Executive Agencies with Respect to the Petroleum Industry," November 8, 1973, especially as it relates to recent fuel shortages. For details on the New England Petroleum Corporation, see 93rd Congress, 1st and 2nd sessions, Senate, Committee on Foreign Relations, Subcommittee on Multinational Corporations, Executive Session, November 27, 1973, p. 29. The *Environment* editorial appeared in Vol. 16, No. 2, "The Oil Glut." Comments from the Joint Economic Committee are contained in United States, Congress, Subcommittees on Consumer Economics, International Economics and Priorities and Economy of Government, "A Reappraisal of US Energy Policy," March 8, 1974.

Chapter 11. Feeble Reform

The quotations on the Federal Energy Corporation proposal are from 93rd Congress, 2nd session, Senate, Commerce Committee, Text and Description of Working Paper No. 1 of the Consumer Energy Act of 1974. For a detailed analysis of taxes and energy policy, see *People and Taxes*, March 1974. It is published by the Tax Reform Research Group, P.O. Box 14198, Ben Franklin Station, Washington, D.C. 20044. The basic federal report on solar energy is entitled "Solar Energy as a National Energy Resource," NSF/NASA Solar Energy Panel, December 1972. The Eggers Committee Report, prepared for the Chairman of the Atomic Energy Commission, is published as the "Report of Subpanel No. 9, Solar and Other Energy Sources," October 1973.

PART THREE: Resistance

Chapter 12. The Georgia Power Project

A description of the project's goals and copies of the newsletter are available from the Georgia Power Project, P.O. Box 1856, Atlanta, Georgia 30301. Joseph Hughes's article was published in *Southern Exposure,* Spring 1973.

Chapter 13. Lifeline Service

For details of the Vermont plan, write to Lee Webb, Goddard College, Plainfield, Vermont 05667. For the Massachusetts plan write Mass PIRG, 233 North Pleasant St., Amherst, Massachusetts 01002. Quotes are from the testimony of Martin L. Puterman, staff consultant, before the Government Regulations Committee, the State House, Boston, Massachusetts, April 2, 1974.

Chapter 14. The Fight against Pacific Gas & Electric

The *Bay Guardian* is the best source of information on Pacific Gas & Electric. J. B. Nielands' article, "How PG&E Robs San Francisco of Cheap Power," appeared March 27, 1969; the grand jury of the City of San Francisco filed one report in December 1973 and an addendum thereto in January 1974. Richard Kaplan's suit, *Charles Starbuck et al. V. San Francisco, PG&E, etc.,* was filed in federal court for the Northern District of California, April 15, 1974. Kaplan's address is 155 Montgomery St., San Francisco, California 94104. Joseph Petulla's article appeared in the *Nation,* August 13, 1973.

Copies of *How to Challenge Your Local Electric Utility,* by Richard Morgan and Sandra Jerabek, are available at $1.50 each from the Environmental Action Foundation, 720 Dupont Circle Building, Washington, D.C. 20036. The Northern Plains Resource Council is based at 421 Stapleton Building, Billings, Montana 59101. The council publishes a newsletter called *The Plain Truth.*

PART FOUR: Proposal for an Alternative Energy System

See *Peoples Appalachia,* Vol. III, No. 1, Spring 1973, available at 1520 New Hampshire Ave., N.W., Washington, D.C. 20036. For a description of TVA's importance in structuring the nuclear power industry, see James Ridgeway, *op. cit.,* p. 14.

Appendix

ABSTRACT OF THE REPORT ON THE INTERNATIONAL PETROLEUM CARTEL BY THE FEDERAL TRADE COMMISSION

Description of the Report.—The Report, which is undated and apparently not finished, consists of 912 pages of text in addition to some 31 statistical tables and 22 charts. It is divided into three parts as follows:

Part I—Resources and Concentration of the World Petroleum Industry
    Chapter I—The World's Petroleum Resources.
    Chapter II—Concentration of Control of the World Petroleum Industry.

Part II—Development of Joint Control over the International Petroleum Industry
    Chapter III—Development of Joint Control over Middle East Oil.
    Chapter IV—Joint Control Through Common Ownership—The Iraq Petroleum Company.
    Chapter V—Other Common Ownerships in the Middle East.
    Chapter VI—Joint Control Through Purchase and Sale of Oil in the Middle East.
    Chapter VII—Joint Control Through Purchase and Sale of Oil in Venezuela.

Part III—Production and Marketing Agreements
    Chapter VIII—Production and Marketing Agreements Among International Oil Companies.
    Chapter IX—Case Studies in the Application of Marketing Agreements in Selected Areas.
    Chapter X—Price Determination in the International Petroleum Industry.

NATURE OF THE ABSTRACT

The statements, conclusions and arguments contained in this summary-abstract come from the Report without any editing, changing or arranging regardless of whether the abstractor thinks them to be justified by the supporting data. Much of the language also comes from the Report even though quotation marks have not been used to so indicate.

THE ABSTRACT

PART I—RESOURCES AND CONCENTRATION OF THE WORLD PETROLEUM INDUSTRY

Chapter I—The World's Petroleum Resources

Petroleum is an exhaustible natural resource; it is of fundamental importance to all phases of industrial activity and indispensable to industrial progress. At times it has been the subject of competition and rivalry; more frequently it has been the subject of agreement and international cartel arrangements. Over the years an international petroleum industry has developed, not only because of the importance of petroleum products, but also because these products are so standardized as to have almost universal acceptance, irrespective of the source of the crude from which they are derived.

World Reserves. As of January 1, 1949, the world's petroleum reserves excluding Russia, were estimated at 73.7 billion barrels, of which more than 90% or 69.0 billion barrels were in 6 countries: United States, Venezuela, Iran, Iraq, Kuwait and Saudia Arabia. The geographical distribution of world reserves is as follows:

	Percent
Middle East	42
North America	37.5
South America	12.8
Russia and its European Satellites	7
Far East	1.3
Western Europe—Less than	0.5

Of the world's total crude reserves, including Russia, American companies control 63%; excluding Russia, 67%.

Most important new discoveries in recent years were in Middle East and Canada. Latter expected to add 1 billion barrels, while Middle East is most prolific potential supplier of oil in the world.

Principal Crude Producing Areas in World.—Crude petroleum is produced in more than 40 countries. In 1949, seven countries—United States, Venezuela, Iran, Saudia Arabia, Kuwait, Mexico and Iraq—produced 85%; if Russia and its satellites are excluded, they produced 92%.

Crude production is concentrated in North America, South America and the Middle East. In North America, three countries, United States, Mexico and Canada, in 1949, produced more than 57% of world's crude.

United States world's principal producer. For several years it averaged about 60%, but in 1949 this declined to 54.7%. In 1949, United States averaged 5 million barrels per day from 449,000 wells, or 11 barrels per well per day.

In 1949, Mexico produced 60 million barrels and ranked seventh. Its daily output per well 160 barrels per day, or 15 times that of United States.

In 1949, Canada produced 59,000 barrels per day, an increase of 174% over 1939 and 81.5% over 1948.

In 1949, Venezuela produced 1,320,000 barrels per day. Per well production was 200 barrels. Same year: Columbia (69 bbls. per well), Argentina (47 bbls. per well, Peru (11 bbls. per well) and Trinidad (25 bbls. per well). Combined production, 240,000 barrels per day.

In 1949, about 15.5% of world's crude output, or 1,400,000 barrels per day, came from Middle East (Iran, Iraq, Saudia Arabia, Kuwait and Bahrein). Per well per day production was, for Iraq, 11,200 barrels. Iran, 2190 barrels, Saudia Arabia, 6,083 barrels, and Kuwait, 4,450 barrels, or an average of 5,143 barrels per well per day for Middle East.

In 1949, the Far East, principally, the Netherlands East Indies and British Borneo, produced about 2% of world's total. This area not yet recovered from war damage.

182 Appendix

In 1949, Russia and Eastern Europe, including Roumania, produced about 8% of world's total, nearly all of which was used in Iron Curtain countries.

Notice that there were wide disparities between distribution of reserves and crude production. In 1949, the United States produced 55% of world's crude from 35.8% of world's reserves, and Venezuela and Mexico also exceeded their percentage of reserves; but the Middle East produced much below its reserve capacities.

The following table shows the pecentage relationship of the world's Crude Reserves and Crude Production, 1949:

(In Percent)

Country	Reserves	Production
United States	35.8	54.65
Kuwait	14.0	2.67
Venezuela	11.5	13.58
Saudi Arabia	11.5	5.16
Iran	8.9	6.08
Iraq	6.4	.94
Netherlands East Indies	1.3	.21
Mexico	1.1	1.80
Total	90.5	85.09
All other countries	9.5	14.91
Grand total	100.0	100.0

World's Crude Oil Refining Capacity.—A country's ability to convert crude petroleum into useable products is indicated by its refining capacity.

The three leading refining centers of the world are the United States, the Venezuela-Caribbean area, and the Middle East.

In 1949, 58 percent of the world's refining capacity was in the United States, about 8 percent in Venezuela and the Netherlands West Indies, and 9 percent in the Middle East. Excluding U.S.S.R. and Eastern Europe, more than 80 percent of the world's refining capacity was located in these three important refining areas. These areas are all important crude producing centers, with the exception of the Netherlands West Indies, which produces no oil but refines large quantities of crude imported from Venezuela and other South American producing countries.

In the Middle East, large refineries are located on the Persian Gulf at Abadan, Bahrein and Ras Tanura. At the eastern end of the Mediterrean are refineries at Haifa and Tripoli, which operate on crude from the Middle East fields. Since the Middle East is not a large consumer of oil, the products from these refineries must necessarily move to other parts of the world—principally Europe and the Far East.

Western Europe held in 1949 approximately 7.5 percent of the world's refining capacity. The total refining capacity of its 82 refineries was less than that of the three large refineries in the Netherlands West Indies. However, since World War II, there has been a tendency to increase the capacity of the European plants. Since Western Europe is not an important producer of crude, practically all of its refiners operate on imported crude. Thus, while total refinery runs to stills in this area were about 522,000 barrels daily in 1949, production of crude in the area was only 30,000 barrels per day.

Exactly the opposite condition exists in South America and the Middle East. There, refinery runs in 1949 were less than daily crude production, indicating exportation of crude.

In 1949, the United States had refinery runs of almost 300,000 barrels per day in excess of crude production. Barring any reduction in crude stocks, the excess of crude runs is an index of the volume of crude imports.

Only a small percentage of the world's refining capacity is located in the Far East. This part of the world is neither an important producing nor consuming area, and refining capacity is about equal to crude production.

World's Petroleum Consumption and Supply.—Some areas of the world consume more petroleum than they produce, and some produce more than they consume.

The only two which produced more than they consumed in 1947 and 1949 were the Caribbean and the Middle East. All other areas, Europe, Africa, the Far East, Oceania and parts of North and South America, consumed more than they produced.

In 1947, the Middle East had an excess of supply over consumption of about 671,000 barrels daily; in 1949, this excess reached 1,191,000 barrels daily, which was almost enough to surpass the Caribbean area as a surplus producer.

In 1949, consumption in the United States exceeded domestic supply by about 322,000 barrels, while in 1947 consumption and supply were about equal.

In 1949, the United States, the Caribbean and the Middle East supplied 85 percent of the world's oil but consumed only 64 percent. Thus a surplus of 21 percent, or 2,120,000 barrels per day, was available for deficit producing countries.

Excluding Russia, the United States is the only important industrialized country which is able to supply its own petroleum needs. Western Europe and Canada (the latter prior to recent discoveries) are highly industrialized but produce little petroleum.

The Pattern of International Trade in Petroleum.—There are four principal movements of crude oil and its products: (1) Crude from producing areas to consuming areas; (2) Crude from producing areas to refining centers which are not consuming areas; (3) refined products from such refinery centers to consuming areas; and (4) refined products from producing areas with refinery facilities to consuming areas.

In 1948, the crude movements from the United States were to Canada, Cuba, Argentina, France and the United Kingdom; from Venezuela and Colombia, large quantities to the Netherlands West Indies, the United States and Europe and small amounts to Africa, Argentina and Uruguay.

In 1948, about 23 million barrels of crude moved from the Middle East to the United States and Canada, or about 12 percent of the total Middle East exports.

In 1948, Venezuela exported 437,700,000 barrels of crude of which 270,992,000 barrels, or 62 percent, went to the Netherlands West Indies, which is neither a producing nor consuming area (it's a refining area). Similarly, movements of crude from the Persian Gulf to Bahrein Island and Palestine are movements to refining centers rather than to consuming areas.

184 Appendix

Since the refined products from these areas must move to consuming centers, the volume of international trade is increased.

In 1948, France had about 50 percent of the refining capacity of Western Europe but its crude production was only 6/10ths of 1 percent of its refinery runs. About 40 million barrels, or 72 percent, of France's crude came from the Middle East and 14.5 million barrels, or 28 percent, from the United States and Venezuela.

Chapter II—Concentration of Control of the World Petroleum Industry

Outside the United States control over the Petroleum Industry (including reserves, production, refining, transportation and marketing) is divided, for all practical purposes, between State monopolies and seven large international petroleum companies, five of which are American and two British-Dutch.

Control by Government Monopoly.—The countries where the state authorities control all phases of the petroleum industry are: Soviet Russia, Albania, Austria, Czechoslovakia, Hungary, Roumania, Poland and Sakhalin. In 1949, these countries controlled 6.1 percent of the world's reserves and 8.4 percent of the world production.

Countries in which petroleum production is under Government operation are: Bolivia, Brazil, Chile, China, Mexico, Spain and Yugoslavia. In 1949, these countries held only 1.2 percent of the world's crude reserves and accounted for 1.8 percent of world production.

The facts about Government operations in the international oil industry appear to be as follows:

1. Russia, Roumania, and Mexico are important producers, but only Mexico has any significant production for the export market.

2. State monopolies for the development and production of oil exist in several other countries, which, however, have limited oil reserves and must import the bulk of their requirements from private corporations.

3. Some state-monopoly countries permit private companies to operate side-by-side with the Government organizations, either as marketers (as in Brazil and Chile) or as integrated producers (as in Peru and Argentina).

4. On an overall basis, state monopoly countries control only about 7 percent of the world's petroleum reserves and 10 percent of world production. But even these figures overstate the importance of Government monopolies, since all state-monopoly countries, except Russia, Roumania, and Mexico, must rely on imports for most of their oil supplies; and, as will be shown later, most of these imports must necessarily come from one or more of the seven international oil companies.

Control by Seven International Petroleum Companies.—The outstanding characteristic of the world's petroleum industry is the dominant position of seven international companies. These seven companies are:

    Standard Oil Company (N.J.): American.
    Standard Oil Company of California: American.
    Socony-Vacuum Oil Company, Inc.: American.
    Gulf Oil Corporation: American.
    The Texas Company: American.
    Anglo-Iranian Oil Company: British.*
    Royal Dutch-Shell Group: British and Dutch.

---

*Fifty-six percent of the stock ownership of this company is held by the British Government.

Control of the industry by these seven companies extends from reserves through production, transportation, refining, and marketing. All seven engage in every stage of operations, from exploration to marketing. The typical movement of petroleum from producer until acquired by the final consumer is through inter-company transfer within a corporate family. Outright sales, arms-length bargaining, and other practices characteristic of independent buyers and sellers are conspicuous by their absence. Control is held not only through direct corporate holdings, by parents, subsidiaries and affiliates of the seven, but also through such indirect means as interlocking directorates, joint ownership of affiliates, inter-company crude purchase contracts, and marketing agreements.

Control of World Crude Reserves.—In 1949, the seven companies owned 65 percent of the world's estimated crude oil reserves; 82 percent of all foreign (outside United States) reserves, 34 percent of the United States (domestic) reserves, and 92 percent of estimated crude reserves outside United States, Mexico and Russia.

Control over World Crude Oil Production.—In 1949, the seven companies accounted for more than one-half of the world's crude production (excluding Russia and satellite countries), about 99 percent of output in the Middle East, over 96 percent of the production in the Eastern Hemisphere, and almost 45 percent in the western Hemisphere. If U.S. production is excluded, their share of the output of the rest of the Western Hemisphere is 80.5 percent. If U.S. production plus that controlled by U.S.S.R. and her satellite countries is excluded, these seven companies accounted for 88 percent of the remaining world's production. This figure understates the degree of control of the private petroleum industry because some of the output outside of the United States and Russia is produced by countries with state monopolies. Since the production of these countries must first satisfy internal demands, it is evident that only a relatively small quantity of crude is available from any source other than one of the seven large companies.

Control over World's Crude Oil Refining Capacity.—In 1950, the seven companies controlled almost 57 percent of the world's crude oil refining capacity. In the Western Hemisphere, excluding the United States, they held more than 75 percent; and in the Eastern Hemisphere, 79 percent. Excluding the United States and Russian controlled capacity, the seven companies owned more than 77 percent of the rest of the world's refining capacity.

Control of World's Cracking Capacity.—Control over cracking capacity is potentially of greater economic significance than control over crude refining capacity. The cracking process enables a refiner to obtain a greater quantity of higher-valued products (for example, gasoline as compared to residual fuel oil) from a given quantity of crude than can be obtained by the straight-run distillation method of refining. The cracking process also enables the refiner to vary the proportions of products produced, thus giving greater flexibility to the refining operation. Moreover, it is only by using cracking processes that the petroleum industry is able to produce high octane gasoline and many of the chemicals which are the basic raw materials for synthetic rubber and many plastics. Thus the cracking process has made the petroleum industry an important supplier of products to other industries, and the concentration of control over the world's cracking capacity thereby affects a broader segment of the world economy than control over crude refining capacity.

In 1950, the seven companies owned 47 percent of the cracking capacity of the United States, 53 percent of that of the Western Hemisphere, 84 percent of that of the Eastern Hemisphere, and 55 percent of that of the world. If the cracking capacity of the United States and Russia (including her satellites) is excluded, the seven companies held 85 percent of all cracking capacity in the rest of the world as compared with 77 percent of the crude refining capacity of the same area.

Control of World's Petroleum Transportation Facilities.—In international trade petroleum and petroleum products are usually transported by tanker. Pipe lines are important in areas where extended movements overland are necessary and feasible; but, outside the United States, this use is limited to moving crude petroleum from producing areas to refineries or water terminals.

As with reserves, production and refining capacity, petroleum transportation facilities outside the United States are also largely controlled by the seven companies.

Tankers.—At the end of 1949, Anglo-Iranian Oil Company and Royal Dutch-Shell, two of the seven companies, controlled approximately 30 percent of the world's tanker tonnage. The other five of the seven, the American companies, controlled more than 20 percent. Thus, the seven companies control at least 50 percent of the world's tanker fleet, and perhaps more, since most of the large oil companies have added substantial tonnage to their tanker fleets in 1950.

If tanker tonnages owned or controlled by governmental agencies are excluded, the percentage of control of the seven companies is about two-thirds of the total privately-owned tanker fleet as compared to one-half of the world's total tanker tonnage.

Pipe Lines.—Outside the United States, every important pipeline in existence or even proposed is controlled by the seven companies, individually or jointly.

Control over Marketing.—While a detailed analysis cannot be given due to the scarcity of statistical information, it would appear that the seven companies are dominant forces in nearly all foreign markets. It does not appear necessary to develop any elaborate statistical argument to support this conclusion. Highly concentrated control over marketing would seem to be inevitable for the simple reason that there are no other companies operating in international markets capable of supplying petroleum products in substantial quantities. Upon the secure basis of their control over production, refining and transportation, the seven companies have built extensive marketing organizations reaching into consuming areas in all parts of the world. The power of these major companies is so substantial as to be virtually unchallengeable, except perhaps, in particular local marketing areas.

Importance of Inter-Corporate Relationships in the International Oil Industry.—The influence of the seven companies on the world's oil business is increased by close corporate relations existing between them.

Joint Ownership of Subsidiary and Affiliated Companies.—Outside the United States, Mexico and Russia, the operations of the seven companies are combined through various inter-company holdings in subsidiary and affiliated companies. These holdings constitute partnerships in various areas of the world. Each of the companies has pyramids of subsidiary and affiliated companies in which ownership is shared with one or more of the other large companies. Such a maze of joint ownership obviously provides opportunity, and even necessity, for joint action. With decision-making thus concentrated in the hands of a small number of persons, a common policy may be easily enforced.

Interlocking Directorates Among the International Petroleum Companies.—Chart 18 shows the interlocking directorates of the companies in the international field. A considerable number of the directors of the seven companies hold multiple directorships in subsidiary companies. For example, it is stated, the directors of the Standard of N.J. and Socony-Vacuum, who determine the policies of Arabian-American Oil Co. (Saudi Arabia) are the same men who help to shape the behavior of the Iraq Petroleum Company. The directors of Anglo-Iranian Company who assist in making high oil policy for Iraq and Iran, participate, along with the directors of Gulf, in planning the price and production policies in Kuwait.

Chart 19 shows some of the indirect interlocking relations between international oil companies operating in the United States and some major domestic companies. These indirect relations provide opportunities for the international companies to harmonize any conflicting interests that might develop between the international companies and major domestic companies. It is an organizational device that could be used in the event it was needed.

Summary of Affiliated and Interlocking Relations.—These seven international companies operate through layers of jointly-owned subsidiaries and affiliated companies. Through this corporate complex of companies, they control not only most of the oil but also most of the world's foreign petroleum refining, cracking, transportation, and marketing facilities. Thus, control of the oil from the well to ultimate consumer is retained in one corporate family or group of families.

Joint ownership of affiliated companies is probably more widespread in the international petroleum industry than in any other field of enterprise. The major international oil companies use the joint-ownership technique not only in conducting foreign operations but also in their operations in the United States and Canada. This is particularly true with respect to control of pipelines and companies holding patents on technological processes. Thus, the international companies, operating in the United States and Canada, are joined with the large domestic oil companies in the two operations where control is likely to exert the maximum of influence on the industry.

Also, the boards of directors that manage the myriad of jointly-owned corporations may, in effect, be private planning boards where differences are resolved and where an oil policy for the world can be established. Under any circumstances, it would be difficult to overlook the significance of the meeting together of directors of the major international oil companies to determine the price and production policies of companies whose operations must inevitably affect the oil industry throughout the world.

Control through the joint-ownership device is further centralized and unified by the fact that directors of the major companies also serve as directors of some of the more important affiliated companies. This close association of policy-making officials can readily result in a unified management of the various combinations of interests, and thus tends to lessen the opportunity for effective competition between the major companies in their foreign operations.

The international companies have also extended their spheres of potential influence over the United States oil industry through indirect interlock-

188 Appendix

ing directorates. Although the association of the directors of the international companies with the directors of important domestic oil companies on the board of a third company may not be significant in and of itself, it at least provides the opportunity for reconciling differences that may arise between the international and the domestic companies.

The significance of this high degree of concentration for the cartel problem lies in the fact that concentration facilitates the development and observance of international agreements regarding price and production policies. Indeed, the concentration of an industry into a few hands may be regarded as the *sine qua non* of effective cartel operations.

PART II—DEVELOPMENT OF JOINT CONTROL OVER THE INTERNATIONAL PETROLEUM INDUSTRY

Chapter III—*Development of Joint Control over Middle East Oil*

Part I of the Report describes the areas of world production and reserves and analyzes the extent of control of reserves, production, refining, transportation and marketing.

Part II gives the details of the growth of concentration in the principal foreign producing areas—the Middle East and Venezuela.

Background.—American oil companies went into the foreign field after World War I for four reasons:

1. Fear of an Oil Shortage in America.
2. High cost of purchasing Private Mineral Rights in America.
3. Discovery of Foreign Reserves.
4. Fear of Foreign Monopoly.

Prior to 1920 American companies either had been indifferent to foreign reserves or they had been largely frustrated in their efforts to acquire reserves in the Eastern Hemisphere, owing to restrictive national and colonial policies of foreign governments and of private oil interests. After 1920, however, they became actively interested in foreign reserves, spurred by the dual fears of a prospective shortage of American oil and of a British-Dutch monopoly of foreign reserves. This increasing interest was also stimulated by the rising costs of American oil and the current and prospective discoveries of large foreign reserves which would afford to their owners a ready supply of cheap oil, advantageously located to important consuming markets.

This growing interest was transferred into successful effort by the American oil companies in acquiring substantial oil reserves in South America, principally in Venezuela, and in the Middle East. The following chapters of Part II describe the process by which these oil reserves were brought under the control of the seven international oil companies.

This control over foreign reserves has been achieved through the use of two techniques, joint ownership and long-term contracts for the sale of crude oil. In the Middle East, the interests of the seven international oil companies have been woven together by joint ownerships of subsidiary companies, each holding interests in one or more of these joint enterprises. This interlacing of control is revealed through the history of the political and private diplomatic negotiations preceding the organization of the jointly-owned subsidiary companies and by an analysis of the management and operational policies of these subsidiaries after their organization. The interests in the Middle East of five of the seven international oil companies have been still more closely interwoven by the execution among them of long-term contracts for the sale of crude oil. The long period of the contracts, the great quantities of oil involved, the unusual nature of the pricing methods and the conditions of sale, and the inclusion of provisions restricting the marketing of the oil suggest that these contracts extend far beyond the ordinary business transaction.

In Venezuela, the three international oil companies, Dutch-Shell Group, Standard Oil Co. (N.J.) and Gulf Oil Corporation, owning the bulk of that nation's oil reserves and production were closely bound together through long-term contracts for the sale of crude oil. Closely allied to these agreements, which in effect bound the three companies together in a partnership were other agreements designed to impose restrictive controls on the production of two of these companies.

Companies Operating in Middle East

I. The following three companies of British nationality control the entire oil reserves in Iraq:

1. *Iraq Petroleum Co., Ltd. (formerly Turkish Petroleum Company).*—75-year concession in provinces of Bagdad and Mosul east of the Tigris River granted in 1925 as a revival and revision of concessions granted by Turkey before World War I.

Stock ownership is as follows:

	Percent
D'Arcy Exploration Co. (British)	23.75
Anglo-Saxon Petroleum Co. (British)	23.75
Compagnie Francaise des Petroles (French)	23.75
Near East Development Co. (Jersey Standard and Socony Vacuum)	23.75
C. S. Gulbenkian (an Armenian-British subject)	5.00

2. *Mosul Petroleum Company (formerly British Oil Development Company).*—75-year concession covering the area west of the Tigris River and north of Lat. 33°, originally granted to British Oil Development Company in 1933.

3. *Basrah Petroleum Co., Ltd.*—75-year concession covering the province of Basrah, granted in 1938.

	Percent
Royal Dutch-Shell Co. (British and Dutch)	23.75
Anglo Iranian Oil Co., Ltd. (British)	23.75
Compagnie Francaise Des Petroles (French)	23.75
Near East Development Co. (American) representing: Standard Oil Co. (N.J.) 11.875% Socony-Vacuum Oil Co. 11.875%	23.75
C. S. Gulbenkian (Syrian individual of British citizenship)	5.00
Total	100.00

II. *Anglo-Iranian Oil Co., Ltd. (originally Anglo-Persian Oil Co., Ltd.).*—Original concession to William D'Arcy in 1901 revised to a 60-year concession in 1933 covering 100,000 square miles in Iran. The company is owned by:

	Percent
British Government	56.0
Burman Oil Co. (British-Royal Dutch-Shell affiliate)	22.0
Individuals	22.0
Total	100.0

III. *Kuwait Oil Co., Ltd.*—75-year concession granted in 1934, covering all of Sheikdom of Kuwait. Company is owned by:

	Percent
Anglo-Iranian Oil Co., Ltd. (British)	50.0
Gulf Exploration Co. (American-Gulf Oil Company)	50.0
Total	100.0

IV. *Bahrein Petroleum Company.*—55-year concession granted in 1940, covering all of Bahrein Islands and territorial waters. The company is owned by:

	Percent
Standard Oil Co. of California	50.0
The Texas Co.	50.0
Total	100.0

V. *Arabian American Oil Co.*—Originally a 66-year concession granted to Standard Oil Company of California in 1933. This concession was augmented in areas in 1939 to cover a total of 440,000 square miles in Saudi Arabia and the entire concession was assigned to Arabian American Oil Co., which is jointly owned by:

	Percent
Standard Oil Co. of California	30.0
The Texas Co.	30.0
Standard Oil Co., N.J.	30.0
Socony-Vacuum Oil Co., Inc.	10.0
Total	100.0

Chapter IV—The Iraq Petroleum Company, Ltd. (IPC)

The Iraq Petroleum Company, Ltd., was the first joint venture in which important American and foreign oil companies were united in one operation. In the words of one of its founders, IPC is a "unique company born of prolonged and arduous diplomatic and economic negotiations. IPC was not organised and operated as an independent corporate entity: rather, its policies and management were determined by and made to serve the mutual interests of the major international oil companies which jointly owned the majority of its shares."

The Report, covering more than 125 pages, sets out, in detail, the history of this company, showing the up-hill fight of the American companies led by Jersey Standard to get a foothold in the Middle East, the bitter quarrels among the partners, British, French, American and the individual Gulbenkian, due to conflict of interests which still persists, the numerous controversies between the company and the Iraqian Government, and the efforts of IPC to keep other oil companies irrespective of nationality out of the Middle East. It makes the charge that the British and American companies deliberately retarded the production of oil in Iraq to the detriment of the Iraq Government, the French and Gulbenkian.

The summary of the Report on IPC is as follows:

The history and development of the Iraq Petroleum Co., Ltd., is a striking illustration of the evolution of joint control through common ownership. By operating through the common ownership [mechanism], the major international companies were able not only to achieve a near-monopoly of oil concessions in a large area of the Middle East, but also to limit production, control prices, and generally restrain competition. As one authority has put it:

"Only groups with world-wide interests and command of proportionate resources could, for instance, afford to bottle up the Iraq production for so many years ... The Largest operator is, more than anyone else, interested in comparative stability, and he will always be prepared to pay for what he thrives on. It is not difficult to appreciate what would have happened to oil markets in general, if the potential production of the Middle East had been unloaded on world markets in the early 'thirties. This is probably the greatest service the major groups—who were all shareholders of the 'Iraq Petroleum'—have rendered to the industry at large."

IPC was not operated as an independent profit-making company. It was essentially a partnership for producing and sharing crude oil among its owners. Its profits were kept at a nominal level as a result of the practice of charging the member groups an arbitrarily low price for crude—a practice which reduced IPC's tax liability to the British Government and permitted the refining and marketing subsidiaries of the groups to capture the major share of the profits resulting from IPC's operations.

Although the origin of IPC dates back to the early 1900's, it did not become important in world oil circles until after World War I. Restrictive arrangements came early in the life of the company, e.g., the Foreign Office Agreement of 1914. For the most part, however, the significant restrictions were not developed until the mid-twenties.

In the early twenties, when the American oil companies first became interested in oil concessions in the Middle East, they placed great emphasis on what was termed the "open-door" policy, and, in fact, made the acceptance of this policy a [*sine*] *quo non* of their participation in IPC. In this, they were actively supported by the American Government. In its initial stages the "open-door" policy was broadly interpreted to mean freedom for any company to obtain without discrimination, oil concessions in mandated areas, particularly in [Mesopotamia]. It was designed to promote active competition among the various companies for oil concessions and to prevent the establishment of a monopoly of oil rights. However, the "open-door" was gradually closed and then "bolted, barred, and hermetically sealed" by a series of deliberate and systematic acts on the part of the owners of IPC. The first of these acts was the concession agreement between TPC (later IPC) and the Iraq Government of March 14, 1925, which made it practically impossible for a nonmember of IPC to obtain a lease or concession in the areas that were to be opened for competitive bidding. Competition for these areas was changed from public or auction bidding to sealed bidding, with IPC given the authority to open the sealed bids and make the awards. In the original "open-door" plan, IPC had been prohibited from bidding on plots to be offered at public auction. This prohibition, however, was omitted from the 1925 concession agreement. Thus, IPC was enabled to outbid any outsider, since (a) under the concession agreement all proceeds from bidding were to go to IPC, and (b) IPC had the right, upon meeting any submitted bid, to award the concession to itself. Secondly, when the groups signed the Red Line Agreement in 1928 and agreed not to be interested in the production or purchase of oil in the defined area (the old Ottoman Empire which included Turkey, Iraq, Saudi Arabia and adjoining sheikdoms, except Kuwait, Israel and Trans-Jordan) other than through the IPC, they "bolted and barred" the "open-door" insofar as their own activities were concerned. Finally, the concession agreement of 1931 closed the door not only on the groups themselves, but on all others as well by eliminating all references to a selection of plots to be offered outsiders, thus giving IPC a monopoly over a large area of Iraq. The "open-door" policy which had been so strongly advanced was discarded in subsequent years without a single test of its adequacy as a practical operating principle.

During the period between 1922, when the "open-door" policy was first advanced, and 1927, when it was in the process of being discarded, radical changes took place in the world oil situation. The fears of an oil shortage which were [so] widespread in 1922 were drowned in a surplus of oil. Instead of competing for the development of oil resources, the international companies turned their attention to limiting output and allocating world oil markets.

With the admission in 1928 of the American group to a share interest in IPC, four of the large international oil companies (Anglo-Iranian, Royal Dutch-Shell, Standard Oil Company (N.J.), and Socony-Vacuum) were united for the first time in a joint venture. The American group, acting as a unit through the Near East Development Corporation (NEDC), along with Anglo-Iranian and Royal Dutch-Shell, comprised the three major groups necessary to control and shape the operating policies of IPC. Following the discovery of oil in Iraq in October 1927, these three groups employed a variety of methods to retard developments in Iraq and prolong the period before the entry of Iraq oil into world markets. Among the tactics used to retard the development of Iraq oil were the requests for an extension of time in which to make the selection plots for IPC's exclusive exploitation, the delays in constructing a pipeline, the practice of preempting concessions for the sole purpose of preventing them from falling into other hands, the deliberate reductions in drilling and development work, and the drilling of shallow holes without any intention of finding oil.

Restrictive policies were continued even after a pipeline was completed, for in 1935, IPC's production was shut back several hundred thousand tons. Moreover, for a time, a sales coordinating committee was established to work out a "common policy regarding the sale of Iraq oil." Again in 1938 and 1939, the Big Three opposed any "enlargement of the pipe line and the corresponding increase in production" on the ground that additional production would upset the world oil market. Although the Big Three eventually conceded to the demands of the French (CFP) for some expansion, no action was taken until after World War II.

An important restrictive feature of the IPC was the Red Line Agreement of 1928, which prevented the member groups of IPC from competing with themselves and with IPC for concessions in an area which included most of the old Ottoman Empire. One writer in commenting upon the Red Line Agreement stated:

"This agreement is an outstanding example of a restrictive combination for the control of a large portion of the world's oil supply by a group of companies which together dominate the world market for this commodity."

Although the strongest proponents of the Red Line Agreement were the French and Gulbenkian, the Big Three were not unalterably opposed to its adoption. The American group, although opposed to the Red Line when first advanced, later agreed to its adoption.

The Red Line gave adequate protection against independent action by the groups within IPC, as was evidenced by IPC's refusal to permit Gulf Oil Corporation from exercising its option to purchase a concession in Bahrein. However, there was a loophole in the agreement in that it did not prevent non-members from seeking concessions within the Red Line area. When an independent organization, the British Oil Development Co. (BOD), obtained a concession in the Mosul area of Iraq, and when another outsider, Standard of California, obtained concessions in Bahrein and Saudi Arabia, IPC began to secure as many concessions as possible within the Red Line area, principally for the purpose of keeping them out of the hands of competitors. To effect the BOD and Standard of California's encroachments in the Red Line area, IPC subsequently obtained control of the BOD concession by secretly purchasing its shares, while the Big Three attempted to come to an understanding with Standard of California regarding the Bahrein and Saudi Arabian concessions.

But the Red Line Agreement proved to be a serious handicap to the Big Three in their efforts to make a deal with Standard of California. It was a handicap to the Big Three because the French and Gulbenkian were unwilling to waive their rights under the Red Line Agreement which entitled them to their pro rata share of any concessions, of any crude produced, or of any products derived from crude produced within the Red Line area. The Big Three tried, either individually or collectively, for almost seven years, to alter the Red Line Agreement in such a way that they would be able to neutralize the competitive effects of Standard of California's operations, with however only partial success. When World War II interrupted negotiations, the IPC groups had reached a temporary understanding among themselves in the form of an Agency Agreement for purchasing Bahrein's production. This agreement, however, fell far short of their real objective, which was a partnership agreement with Standard of California and Texas Co. covering the Arabian concession.

During the war, the Red Line Agreement was more or less put aside. Some of the groups, like CFP and Gulbenkian, were considered enemies and could not share in IPC's production nor actively participate in the management of the company. Others, like Standard Oil Company (N.J.), could not lift their share of crude because of shipping restrictions. Despite the disruptions of the war, IPC followed its established price policy which assured a low price to the groups able to take IPC crude, although such a policy was of course unfavorable to the interests of the inactive parties.

At the end of the war, CFP and Gulbenkian were reinstated and the question of reaffirming the Red Line Agreement became a pressing issue. The American group, instead of reaffirming the Agreement declared it dissolved because (a) some of the owners had been enemies during the war, and (b) its restrictive provisions violated the American antitrust laws. It must also be remembered that at this time, Jersey Standard and Socony-Vacuum (the American group in IPC) were extremely anxious to purchase an interest in the Arabian American Oil Co. (Aramco), the company which held the Standard of California and Texas Co.'s concession in Saudi Arabia. It was the Red Line Agreement which before the war had blocked their efforts to secure a participation in these same concessions, the value of which had increased immensely during the war. It not only had been proved, but had developed into one of the world's most important oil concessions. As long as the Red Line Agreement hung around their necks like a millstone, the Big Three were placed in the role of unwilling outsiders, watching Standard of California develop this great new area, with possible disastrous effects on world price markets.

Following declaration by the American group that the Red Line Agreement was dissolved, the French (CFP) filed a suit in the British courts to enforce the Agreement and to obtain their proportionate share of any interest which the American group might secure in Aramco. But in November 1948, before the court case came to trial, a new agreement was negotiated freeing the IPC groups from many of the restrictive provisions of the Red Line Agreement and permitting the American group to purchase an interest in Aramco. The new agreement made it somewhat easier for a member of IPC to obtain oil in excess of its pro rata share. This was of considerable benefit to CFP, which for years had desired more oil from IPC, but under the Red Line Agreement was permitted to take only its pro rata share of IPC's production, as determined by the majority of the groups.

Under the new agreement, the groups were free to engage not only individually but also in common ownership arrangements with other sections of the Red Line area. The groups continued to obtain crude at an arbitrarily low price, and the Big Three retained their position of control over IPC's policies and management. That its long-established policy of restrictionism still continues is suggested by a proposal to restrict IPC's 1950 drilling budget to actual requirements; this proposal was advanced because of a fear that the development of any additional producing capacity would enable CFP to bargain for higher production rates.

The major groups in IPC not only restricted IPC's output, but after World War II, when the exigencies of the moment made it necessary for IPC to operate a refinery at Tripoli, they concertedly acted together to control the prices and the distribution of the refinery's products in such a manner as to discriminate against outsiders and further the interests of the major groups' marketing organizations. Also, the prices established by IPC for products at the Tripoli refinery, as well as the prices charged consumers in Iraq, ignored their natural economic advantage of location near a low cost source of supply.

In summary, the fundamental purposes and objectives of IPC were described by the French in a confidential document:

"The incorporation of IPC and the execution of the Red Line Agreement marked the beginning of a long term plan for the world control and distribution of oil in the Near East."

IPC was so operated as: ". . . to avoid any publicity which might jeopardize the long term plan or the private interests of the group . . ." It would appear there is no evidence that this "long term plan" is still not in effect.

Chapter V—Other Common Ownerships in the Middle East

In addition to IPC, other joint ventures of major importance in the Middle East are the Arabian American Oil Co. (Aramco) and the Kuwait Oil Co., Ltd. The combined average daily crude production of these companies, in 1950, was approximately 900,000 barrels which was more than one-half of the Middle East total production.

The Arabian American Oil Co.—Aramco has extensive oil operations, and, except for the Bahrein Petroleum Company's operations on Bahrein Island in the Persian Gulf, is the only company holding an important oil concession in the Middle East that is exclusively American owned and operated.

In 1950, Aramco accounted for 5.3% of world crude production and about 35% of all production in the Middle East. Although Aramco did not discover oil in Saudi Arabia until 1938, production increased from 11,000 barrels per day in 1939 to 547,000 in 1950, which made it the second largest producer of oil in the Middle East, second only to Anglo-Iranian in Iran.

Aramco is now owned by The Texas Co. (30%), Standard Oil Company of California (30%), Standard Oil Co. of New Jersey (30%) and Socony-Vacuum Oil Co., Inc. (10%).

Kuwait Oil Co., Ltd.—Gulf Exploration Co., a subsidiary of Gulf Oil Corporation, and Anglo-Iranian Oil Co., a British corporation in which the British Government owns 56% of the stock, each owns a 50% share of the stock of Kuwait Oil Co., Ltd., an operating oil company registered under British law. The Kuwait Co. holds an exclusive concession to explore for and produce oil in the entire 6,000 square miles of territory constituting the independent Sheikhdom of Kuwait on the west coast of the Persian Gulf. The Sheikh of Kuwait rules this area under British protectorate and has treaties and agreements with Great Britian giving preference to British subjects or companies respecting exploration for and production of oil in Kuwait. The Kuwait concession is for 75 years beginning on December 23, 1934 and covers about 3,900,000 acres. The one field so far developed at Burgan has been estimated by Gulf to contain 10 billion barrels of oil of which Gulf will get half. In 1946, Gulf estimated that in other areas in Kuwait it controlled something less than 2 billion barrels.

A summary of the background and development of these two companies, Aramco and Kuwait, is as follows:

When Standard Oil Company of California and Gulf Oil Corp. first began negotiations to obtain concessions in Saudi Arabia, both were practically newcomers in the world oil trade and both were acting outside the closely cooperating group of international oil companies that controlled Middle East production.

The older companies first tried to prevent Gulf and Standard of California from obtaining concessions, and when this failed attempted to devise other means of preventing the newly-discovered oil from disturbing world markets. The history of the development of Aramco and Kuwait Oil Company, Ltd., therefore, is a history of the difficulties faced by independent American interests in obtaining a foothold in the Middle East and of finding a market for flush production.

196 Appendix

Aramco.—In winning an exclusive concession covering an area of about 360,000 square miles in Saudi Arabia, Standard Oil Company of California faced the competition of Iraq Petroleum Company representing the combined interests that controlled all production in Iraq and Iran, as well as in most of the remainder of the Middle East. To obtain the concession, California Standard agreed to make loans and advances to the Arabian Government in the amount of £150,000: to pay annually a cash rental of £5,000; and if oil were discovered, to pay a royalty of 4s. per ton and to furnish free to the Saudi Arabian Government 200,000 gallons of gasoline and 100,000 gallons of kerosene, annually. Further advances of £100,000 and an increase in the annual rental of £20,000 was the price for obtaining a second concession in 1939 which extended the concession area to 440,000 square miles. The loans and cash advances were recoverable by the company only by deductions from royalties, and the government also hoped to augment its income through royalties. From the outset, therefore, Aramco's financial relationships with the Saudi Arabian Government required a market outlet for any oil discovered.

With no established position in the Eastern Hemisphere, Standard of California had already come face to face with the difficulty of finding a market for its Bahrein oil without engaging in a competitive struggle with the established international companies. The Bahrein problem was solved on July 1, 1936 when Standard of California bought a half interest in the Far Eastern marketing facilities of the Texas Company, which already had an established position east of Suez, with the Texas Company buying a half interest in the Bahrein Petroleum Company. Aramco also obtained accession to these limited eastern markets when the Texas Company, in December 1936, bought a half interest in Aramco. A press comment at the time of the first acquisition stated that: "* * * it assures that Bahrein production as well as any output that may eventually come from countries now being developed by Standard Oil Company of California will have assured and regulated outlets and will so lessen any possible danger of upsetting the equilibrium of international markets."

Up to 1941, California-Texas Oil Company (Caltex), the marketing company owned jointly by Standard of California and Texas, was able to [find] markets east of Suez for only 12,000–15,000 barrels daily of Aramco's oil. This was reported to be less than one-seventh of what Aramco's developed fields could have produced in 1941. During the war, production was gradually increased to 58,386 barrels daily in 1945. Since a large proportion of this output, however, was refined and sold to the Allied Governments, this proved to be only a temporary outlet, leaving Aramco at the end of the war with crude oil and refining facilities, but no market. Moreover, Aramco's need for markets was aggravated by the discovery of additional fields in 1945 and 1947. At this point, Aramco proposed to build a pipeline to the Mediterranean.

This proposal caused great concern to the established international companies, which immediately endeavored to open up additional markets to Aramco, both east and west of Suez, but in such a manner as not to disturb world markets. This involved several coordinated [steps]. First, the Texas Company sold its European marketing facilities to Caltex, thus making its market *west* of Suez available to Aramco. Second, Standard of California and Texas permitted Standard Oil Company (N.J.) and Socony-Vacuum Oil Company, together, to purchase a 40-percent interest in both Aramco and

Trans-Arabian Pipe Line Company. And third, Jersey Standard and Socony-Vacuum entered into contracts to buy oil from Aramco. Thus, while new markets were opened up to Aramco, the recognized marketing positions of the international oil companies were preserved. The principal change was a shift in their sources of supply on the part of three of the four American companies which now own Aramco in order to make room for Aramco's production—production which they are now in a position to control.

Kuwait Oil Company.—The history of Kuwait Oil Company, Ltd., is likewise one of an American company which, single-handed, sought to obtain a foothold in Middle East production. In 1931, both Eastern Gulf Oil Co., a subsidiary of Gulf Oil Corp., and Anglo-Persian, separately, began negotiating for a concession in Kuwait. After about 3 years, during which time oil was discovered in nearby Bahrein, Anglo-Persian and Gulf made common cause to obtain an exclusive concession covering the whole of Kuwait, setting up Kuwait Oil Company, Ltd., on a 50/50 basis to operate the concession.

At the insistence of Anglo-Persian, the contract establishing the operating company contained a provision that neither party would use oil from Kuwait to upset or injure the other's "trade or marketing position directly or indirectly at any time or place," and that they would confer from time to time to settle any questions that might arise between them regarding the marketing of Kuwait oil. In addition, the contract provided that the quantity of oil to be produced in Kuwait would consist of two parts: (1) such quantity as the two owners agreed to produce and share equally, and (2) such additional quantity as either party might order out for its own account. If, however, Gulf called for additional quantities, as provided in (2), Anglo-Persian reserved the right, at its sole option, "to supply Gulf's requirements from Persia or Iraq in lieu of requiring the company to produce oil or additional oil in Kuwait." Thus, Anglo-Persian reserved the right to control the quantity of oil produced in Kuwait by substituting oil from its other sources.

Although oil was discovered in Kuwait in 1938 and further explorations proving the existence of large reserves continued until 1942, lack of transportation and loading facilities prevented the sale of oil until after the war. The necessary facilities were completed and commercial sales began in August 1946, with Gulf selling its share of Kuwait's production to a Royal Dutch-Shell subsidiary. Nine months later, Shell contracted to take much larger quantities of Gulf's Kuwait oil for a long period of years. The manner in which much of Gulf's share was fitted into the world market under this commercial agreement without disturbing competitive positions is discussed in the next chapter.

Chapter VI—Joint Control through Purchase and Sale of Oil in the Middle East

Throughout the world the "Big Seven" oil companies buy and sell among themselves, through contracts to purchase and sell, large quantities of both crude and refined products.

When these purchase agreements are discussed publicly by representatives of the petroleum companies, emphasis is usually placed on the ordinary commercial purchase and sale aspect which they share with all other sales contracts. However, since the companies participating in them often are already bound together through joint-ownership arrangements and participate together in various production and marketing agreements, purchase and sale contracts among them often lack many of the arms-length features that characterize ordinary commercial agreements among mutually independent buyers and sellers. Under these circumstances, the sales of oil covered by the contracts can often be utilized as an instrument to divide production, restrain competition in marketing, and protect the market positions both of the buyer and the seller. They determine who may or may not buy crude oil from particular producing properties. They tend to funnel the production from more or less diversified ownerships into the centralized marketing organizations of the large companies. They tend to keep surplus supplies of crude oil out of the hands of independent oil companies. The existence of these contracts in an atmosphere of joint ownership of production and marketing, the long periods for which they run, the manner in which prices are determined under them, and the marketing restrictions often written into them, indicate that they are something more than ordinary commercial purchase and sale contracts.

The contracts so characterized above, and discussed in this chapter, are:
1. Gulf-Shell Agreement of 1947.
2. Anglo-Iranian Agreements with Jersey Standard and Socony-Vacuum for Sale of Crude Oil (1947).
3. Jersey Standard-Anglo Iranian Sale of Crude Oil Contract (1946-1947).
4. Socony-Anglo Iranian First Purchase Agreement (1947).
5. Socony-Anglo Iranian Second Purchase Agreement (1949).
6. Middle East Pipelines Agreement (1948).
7. Supplementary Agreements of April 5, 1949.

Each of the contracts are discussed in detail. A summary of the discussion is as follows:

The above-mentioned contracts for the sale of crude oil represent still another intermingling of the interests of the major international oil companies in Middle East oil. Joint ownerships in the Middle East, which have resulted in extensive controls and restrictions on production, have been described in Chapters IV and V. In addition to joint ownerships, the crude oil supply contracts described in this chapter have provided another basis for joint control over oil production and marketing. These contracts have resulted in a sharing of oil production in Kuwait and Iran and a channeling of the oil to the market in the hands of firms able and interested in maintaining world prices and markets. These two instruments of control utilized in the Middle East, joint ownership and crude oil supply contracts, have, in effect, served to complement each other in protecting the mutual interests of the international oil companies in the production and marketing of world oil.

The contracts provide for the sharing of large quantities of oil over a long period of time. Under the Gulf-Shell contract, Shell acquired control over 1¼ billion barrels of Kuwait oil owned by Gulf to be delivered over an open-end contract period of at least 12 years. Under the Anglo-Iranian—Jersey Standard-Socony contracts, Anglo-Iranian turned over to the two American companies 1.3 billion barrels of Kuwait-Iran oil over a 20-year contract period. Thus these contracts result in the division of the production of Kuwait and Iran between the buyers and the sellers and, in effect, [gives] them mutual and continuing interests in that production over a period of many years.

These mutual interests are typified by the unusual terms as to price that were agreed upon by the parties. Under the Gulf-Shell contract, no price is stated, but elaborate provisions were written providing for the division of profits between the two parties. The profits are determined and shared for the entire integrated process of producing, transporting, refining, and marketing for a minimum period of 12 years. Thus, to all intents and purposes, Gulf and Shell are joined together in a long-term integrated oil enterprise.

A "cost-plus" pricing principle was adopted for the three contracts for the purchase of oil owned by Anglo-Iranian. The price under the Jersey Standard and Socony First Purchase Agreements was fixed at the actual cost of production plus a fixed sum of money per ton, and under the Socony Second Purchase Agreement, at actual cost plus one third of the gross profit per ton realized on the crude oil. Such a pricing principle gives the purchaser a direct and strong interest in the costs of the seller, since the purchaser will benefit from any economies achieved by the seller in his operations. This interest was evidenced by the extensive provisions in the contracts setting forth the method for determining and allocating costs and for the delivery by Anglo-Iranian of "any and all information" relating to the cost elements entering into the price which Jersey Standard and Socony might "reasonably" request.

The significance of the contracts as instruments for the control of Middle East oil is further [evidenced] by the provisions restricting and controlling the marketing of the oil. Under its 1933 joint-ownership agreement with Anglo-Iranian, Gulf was restricted as to the markets which it could enter and, in any case, was bound not to disturb Anglo-Iranian's marketing position at any time and at any place. This restriction, which particularly restrained Gulf from entering markets "east of Suez," was carried forward in the Gulf-Shell agreement, and detailed instructions were added, specifying the market territories in which Shell could distribute the oil and the marketing organizations through which the oil would be sold. The specified marketing organizations included many companies jointly owned by Shell and Anglo-Iranian, as well as marketing subsidiaries of Shell. The effect of the restrictions on marketing in the 1933 joint-ownership agreement and in the Gulf-Shell contract was to limit Gulf to those markets in which *it* held a historic marketing position but to allow it, through the profitsharing arrangement, to participate in the marketing of oil in other territories in which *Shell* held a marketing postion. The extensive joint marketing arrangements of Shell and Anglo-Iranian assured the integrity of the marketing positions of both. Thus the effect of these arrangements was to carry forward into the postwar period, the "as is" principle, to be described in Chapters VIII and IX, preserving the historic positions of participants in each market.

The areas in which Jersey Standard and Socony could dispose of the large quantities of oil acquired from Anglo-Iranian were similarly specified in the contracts. Not more than 5 percent of this oil could be distributed "east of Suez," a provision penalizing any excess shipments to this area being inserted in each of the Supplementary Agreements to go into effect on January 1, 1952. Jersey Standard and Socony, under the First Purchase Agreement, were to distribute their oil in Europe and North and West Africa. Socony, under the Second Purchase Agreement, was to import its oil into the United States. In short, under these agreements, the three parties agreed upon the markets into which this oil was to flow.

Thus the crude oil supply contracts, not only because of the large quantities of oil and the [long periods] of time that were specified, but also because of the unusual provisions as to price and marketing, constitute effective instruments for the control of Middle East Oil. As such, they complement and increase the degree of joint control over Middle East oil resulting from the pattern of Joint ownership described in the preceding chapters. The operation of these two instruments of control, in effect, brings the seven international oil companies, controlling practically all of the Middle East oil resources, together into a mutual community of interest.

Chapter VII —Joint Control Through Purchase and Sale of Oil in Venezuela

The use of crude oil supply contracts to control the production of oil had effectively bound together the major oil interests in Venezuela a decade before this instrument of control was adopted in the Middle East. In Venezuela, however, such contracts were of a far more comprehensive nature, welding the interests of the parties together in an explicit joint enterprise lasting for the life of the concessions owned by the joint enterprise. Under this agreement the price paid by the purchasers, except for the sum initially paid as a consideration for the agreement, was merely the actual costs of production of the oil. The contracts, therefore, were again devices for sharing the ownership of the oil. Another important feature of the contracts was certain controls that were laid on production.

The Venezuelan petroleum industry has been an important factor in world petroleum markets for about 25 years. Throughout this quarter of a century of almost uninterrupted growth, three great international petroleum companies—Royal Dutch-Shell, Standard Oil Company (N.J.), and Gulf Oil Corporation—have been closely associated in the exploitation of this rich Venezuelan resource. These three companies have jointly maintained a pervasive control and influence over the Venezuelan industry in all its aspects, from exploration and development to the marketing of the end products.

During the period before World War II, practically all of Venezuelan petroleum had been produced in western Venezuela. However, in the mid-1930's a mounting tide of discoveries indicated that a resource of unpredictable magnitude and richness existed in eastern Venezuela. While practically all of these newly-discovered oil fields were held by subsidiaries of Standard (N.J.) and Gulf, it was evident that the impact of the new production upon world petroleum markets would concern all of the international oil companies, particularly Standard and Shell. Accordingly, the various subsidiaries of Standard, Shell, and Gulf entered into agreements designed to attain the following objectives:

1. The virtual elimination of Gulf as an independent factor in Venezuela. This was accomplished by the transformation of the Mene Grande Oil Company, Gulf's operating subsidiary in Venezuela, into a joint enterprise, owned and controlled by Gulf, Shell, and Standard (N.J.), and by the attendant surrender by Gulf of valuable management prerogatives. Two so-called sales of oil agreements were the principal medium for this transformation.

2. The control and regulation of petroleum products of all Venezuela—eastern and western alike—so that Venezuelan output would, at all times, accord with the current world market situation as seen by the producing companies. A production quota system was set up to achieve this goal.

3. The control and regulation of the development of the newly discovered eastern Venezuelan oil fields. This was accomplished partly by the re-creation of Mene Grande as a joint enterprise, partly by the production quota system, and partly by the merging of a major portion of the eastern Venezuelan holdings of Mene Grande and of Standard (N.J.) into a joint enterprise.

The agreements were designed to meet a specific situation and to gain specific ends. They were signed in December 1937, a year which is thus of critical importance in the analysis of the Venezuelan oil industry.

IDENTIFICATION LIST OF COMPANIES

The principal *operating* companies in Venezula in 1937, with identifying abbreviated names, were as follows:

a. Gulf Oil Company subsidiary Mene Grande Oil Company, C.A. (Meneg).

b. Standard Oil Company (N.J.) subsidiaries:
Lago Petroleum Company (Lago).
Standard Oil Company of Venezuela (SOV).
Creole Petroleum Company (Creole).

c. Royal Dutch-Shell group subsidiaries: Venezuelan Oil Concessions, Ltd., Caribbean Petroleum Company, Ltd., Colon Development Company, Ltd. (Shell).

In addition, two companies whose principal activities were in other countries entered into the Venezuelan picture by virtue of participation in the agreement. These were:

d. Standard Oil Company (N.J.) subsidiary: International Petroleum Company, Ltd. (International).

e. Royal Dutch-Shell group subsidiary: N. V. Nederlansche Olie Maatchippij (NOM).

The reason for the three major oil groups entering into a series of agreements in 1937 was because, in that year, Venezuela was exporting to the United States only 28% of its crude oil production. The balance was, therefore, thrown on the world market (outside the United States) in which Jersey, Gulf & Shell had large interests, particularly the Royal Dutch-Shell group which accounted for 22.4% of all crude production outside the United States.

The principal agreements considered in this chapter are: (1) The Meneg-SOV Agreement of December 15, 1937; (2) The Meneg-International (Principal) Agreement signed in Toronto, Canada, on December 15, 1937; (3) The International-NOM Agreement signed on November 30, 1938 but made retroactive to December 15, 1937; (4) The four-Party Ratio Agreement, the most important quota agreement, signed on December 15, 1937 (later cancelled as of December 31, 1942); and (5) The International-NOM Ratio Agreement, also signed December 15, 1937. There were also some side or collateral agreements in connection with one or more of the principal agreements.

SUMMARY OF THE SIGNIFICANCE OF THE AGREEMENTS

As of December 15, 1937, the Mene Grande Oil Company, a 100-percent subsidiary of the Gulf Oil Company, owned various oil concessions in Venezuela in its own right and owned interests in various other concessions jointly with others. It owned and held interests in the ownership of physical properties in connection with these concessions and concession interests. It held the right to develop and to participate in the development of them. It owned rights to future oil production from these concessions proportionate to its ownership. The net book value of these concessions, physical properties, and rights was $19,562,852.

On December 15, 1937, Meneg participated in a number of agreements with various Standard (N.J.) subsidiaries. The net effect of these agreements and of subsequent agreements between the International Petroleum Company and the NOM, a Shell subsidiary, may be summarized as follows:

a. Meneg and SOV, a Standard (N.J.) subsidiary, agreed, as of December 15, 1937, to "pool" all the concessions in which they had joint interests. These "pooled concessions" represented about 74 percent of all exploitation concession interests and about 90 percent of all exploration concession interests held by Meneg. Practically all of these were located in eastern Venezuela. While the concessions were to be "operated" by their legal "owners" the actual management and control over the exploration, development and production was transferred to a committee representing equally the two partners. Additional concessions could be "pooled" at the discretion of and upon agreement of the committee. All costs of operation of each "pooled concession" were to be joint and equally-shared costs and one-half of the oil produced on each "pooled concession" was deliverable to each of the parties. Thus, a large part of the properties of the two companies was, in fact, transferred from the separate management to that of a joint enterprise.

b. Meneg sold an undivided one-half interest in all of its concessions, physical properties, and rights, as defined in the opening paragraph of this summary section to the International Petroleum Company, an indirect subsidiary of Standard (N.J.), the sale taking effect on December 15, 1937. Included in this sale was an unlimited and continuous option to International to purchase all concessions and concession interests that Meneg might want to sell, transfer, or surrender, and a similar option to buy an undivided one-half interest in all new concessions and concession interests that Meneg might acquire. Included in the sale, also, was an undivided one-half interest in all physical properties of every description acquired by Meneg after the date of this sale. International later sold to a subsidiary of Royal-Dutch Shell, NOM, retroactively to December 15, 1937, an undivided one-half interest in all the concessions, properties, rights and options it had acquired from Meneg.

As of and after December 15, 1937, therefore, all of the assets, tangible and intangible, of Meneg were vested as follows: Gulf, 50 percent; Standard (N.J.), 25 percent; Shell, 25 percent. Insofar as the "pooled concessions" were concerned, as distinct from all of Meneg's other assets, their ownership and all oil produced therefrom was divided as follows: Gulf, 25 percent; Standard (N.J.), 62.5 percent; and Shell, 12.5 percent.

c. Meneg surrendered valuable management prerogative to its partners. (1) Meneg granted to International the right of access to and inspection of all of its properties, records, books of account, and all information, including geological, in its possession relating to its concessions and properties. (2) Meneg also agreed to submit annually budgets of capital and operating expense and statements of operating policy for all its properties and interests for the approval and concurrence of International, such approval being *necessary* for all exploration and development work other than that in connection with current and "normal" operation. (3) Meneg and Standard (N.J.) operating subsidiaries agreed to fix total production schedules beginning in 1938, quotas being fixed under these schedules in the ratio, respectively, of 100:345. Machinery was provided for a revision of these production schedules and ratios beginning in 1950. The superior and dominating position of the Standard (N.J.) subsidiaries was recognized by Meneg. These provisions were to be read and interpreted together with those relating to budgets and operating policy, i.e. Meneg's limited voice in the quota arrangement was subject to approval by International. While the quota arrangements were cancelled in 1943, the budgeting arrangements are still in effect.

International admitted Shell, through NOM, to participating on a substantially equal basis in all the supervision rights it had acquired under the provisions summarized above. NOM thus acquired an equal right of access and of management through International. Similarly, it acquired knowledge of and approved and consented to the production quota arrangements. Meneg, in two separate documents, recognized International's right to admit NOM to these arrangements and to the sale.

d. In consideration of the sale and transfer of prerogatives as described above, International agreed to assume one-half of all of Meneg's "current obligations" arising after December 15, 1937, or allocable to oil produced after that date, and to pay Meneg $100,000,000. The consideration paid by NOM to International for the sale of one-half of these acquisitions included the assumption by NOM of one-half of all of International's burdens, plus an additional payment by NOM to International of $25,000,000. The net effect of the various considerations paid may be summarized as follows:

(1) The agreement by the parties to assume varying portions of "current obligations" represented, in effect, an agreement to pay all the costs of oil delivered and to be delivered to them after December 15, 1937. The payments were to be proportionate to the interest in the oil produced, i.e. Gulf, 50 percent; Standard (N.J.), 25 percent; and Shell, 25 percent. The total amounts paid by the three parties from December 15, 1937 to October 31, 1947, a period of just under ten years, were as follows:

Gulf (50%)	$172,033,708
Standard (N.J.) (25%)	86,016,854
Shell (25%)	86,016,854
Total	344,067,416

Meneg's net production of oil during the same period was similarly divided:

	Barrels
Gulf (50%)	191,073,259
Standard (N.J.) (25%)	95,536,630
Shell (25%)	95,536,630
Total	382,166,519

(2) As of December 15, 1937, the three parties had the following investments in Meneg's assets and rights: (a) Gulf owned an undivided one-half interest in the enterprise (net book value $9,781,426) and had a net "profit" on

the sale of the other half of $90,218,574, i.e. it held its one-half interest free of cost, plus cash in hand or owed to it of $80,437,148. (b) Standard (N.J.) owned an undivided one-fourth interest (net book value, $4,890,713) at a cost of $25,000,000. (c) Shell owned an undivided one-fourth interest (net book value, $4,890,713) at a cost of $75,000,000.

These investments were in addition to and separate from the burdens assumed and carried by the three parties for "current obligations," i.e. for the costs of producing oil after December 15, 1937. In accounting practice they represent charges paid for the rights to future production of oil from the concessions and properties owned by Meneg and the right to participate in such growth and expansion as Meneg might enjoy in the years to come.

At the time the various agreements were signed, Standard and Shell were both much larger producers in Venezuela than Gulf. It was a time when the world petroleum industry was greatly concerned with the problem of "overcapacity"—a problem which could have been greatly aggravated by uncontrolled Venezuelan production. The interests of Standard and Shell lay in "stabilizing" and controlling Venezuelan production in relation to markets for petroleum and petroleum products elsewhere in the world. The various agreements entered into facilitated the achievement of this goal, and the exchange of money and rights to oil was the instrument for fixing the interests of the parties. For cash Gulf surrendered its management prerogatives and accepted a position as a junior partner in the Venezuelan industry. Standard (N.J.) and Shell consolidated their control over the Venezuelan petroleum industry. Production quotas were fixed for about 70 percent of Venezuelan capacity. Shell alone remained outside of the quota arrangement but had knowledge of and consented to it.

These contracts for the sale of oil, therefore, had unusual features that distinguished them from ordinary business transactions. The three parties to the contracts became in reality partners in a joint enterprise, each holding undivided interests in the assets of the joint enterprise. The "price" paid by Jersey Standard and Shell for the oil they received was the cost of production of that oil. The "profit" accruing to Gulf for the "sale" of oil was the fixed sum initially paid, i.e. $100,000,000 less one-half of the book value of Meneg's assets in 1937, irrespective of the quantities of oil that would actually be delivered over the life of Meneg's concessions and interests in Venezuela. The "sale of oil" features of these contracts thus appear actually as a sharing of oil among the three companies, while the contracts themselves appear to have been designed to further the regulation and control of the development of crude oil production in Venezuela.

Chapter VIII—Production and Marketing Agreements among International Oil Companies

In addition to (1) their outright ownership of reserves and production and distribution facilities, (2) their joint ownership arrangements, as in the Middle East, and (3) their agreements on the exchange of crude, the leading oil companies have also exercised control over the industry through restrictive agreements regarding international production and marketing. These agreements, some of which are the subject of this chapter, constitute a series of steps by which the major international companies sought to establish more effective control over distribution and prices throughout the world.

A feeling that greater control in distribution markets was needed arose in the middle 1920's. At that time it became evident that existing degree of control over production constituted an inadequate basis for controlling price competition, even among the largest companies: for a price war broke out in Asia between two of the large international interests, Standard of New York (new Socony-Vacuum) and Royal Dutch-Shell groups, and thereafter, in 1928, an international cartel agreement was formed among the principal international companies. Its purpose was to forestall market competition, both among themselves and with others, through control of production and exports. But this agreement left largely uncontrolled the activities of important independent interests that were not parties to the agreement. It was followed by other agreements designed, first, to forestall competition more fully among the parties to the main agreement and, second, to extend the international groups' control over the activities of independents through separate local marketing agreements formulated so far as possible, in accord with the principle and procedures agreed upon by the parties to the main agreement. These agreements covered the period from 1928 to the outbreak of World War II. They were an integral part of the pattern of concentrated big company control over world oil resources and trade described in other chapters of this Report.

The principal international oil agreements between 1928 and 1934 which are discussed in this Chapter are:

1. *The Achnacarry Agreement,* dated September 17, 1928. The title on this agreement as finally adopted is "Pool Association." It is also referred to as the "As Is Agreement of 1928."

2. *Memorandum for European Markets, 1930.* This is dated January 20, 1930.

3. *Heads of Agreement for Distribution,* adopted December 15, 1932.

4. *Draft Memorandum of Principles—1934,* adopted in May-June 1934.

The primary movers in the four great international agreements, dated from 1928 to 1934, and in efforts to implement them, were the three major international groups of the oil industry, namely, the Standard Oil Co. (N. J.) group, the Royal Dutch-Shell group, and the Anglo-Iranian Oil Co., Ltd., (formerly Anglo-Persian Oil Co., Ltd.) group. Each of these is a completely integrated oil group consisting of a central controlling company and its subsidiary and affiliated producing, refining, transportation and marketing interests. In addition, each of these groups is associated in interest with each of the others and, in some instances, with outside interests through joint ownership of reserves, producing and refining capacity, and marketing facilities, and through agreements for the purchase and sale, exchange, or marketing jointly of crude oil and refined products.

The immediate background for the first international agreement in 1928 was a price war between Royal Dutch-Shell and Standard Oil Co. of New York in India in 1927, which promptly spread to the United States and European markets, where it severely affected, or threatened to affect, the financial interests of all major and minor companies. After this controversy was settled, the heads of the three major groups met and negotiated the Archnacarry Agreement of 1928.

This agreement was essentially a declaration of principles and procedures by which the three majors hoped to forestall in the future similar outbreaks of competition, both among themselves and with others. The heart of this agreement was a group of seven principles to govern group action. These principles, collectively called "as is", were:

(1) Acceptance by the participants of their "present [1928] volume of business and their proportion of any future increase in consumption."

(2) Joint use of existing facilities.

(3) Construction of only such additional facilities as were necessary to supply increased demand.

(4) Production to retain the advantage of geographical location on the basis that "values of products of uniform specifications are the same at all points of origin."

(5) Supplies to be drawn from the nearest producing area.

(6) Excess of production over consumption to be shut in by producers in each producing area.

(7) Elimination of any competitive measures or expenditures which would materially increase costs and prices.

These principles were never formally abandoned. On the contrary, in subsequent agreements intended to implement them in whole or in part, "as is" was said to be of such a durable nature that failure of the subsidiary agreements to accomplish their objectives, or even their total abrogation by some parties, should not prevent further joint effort to find a basis on which to make "as is" effective.

In 1928, the world's principal developed reserves and production, outside of Russia and the United States, were largely controlled by the three participants in the Achnacarry Agreement. The basic assumption underlying this agreement was that world production might be fitted to consumption and world prices thereby controlled if these companies (a) by agreement limited production and exports from the areas which they controlled; (b) found a way lawfully to control exports from the United States, where diverse ownership and the antitrust laws made direct control of production by agreement impracticable, and (c) found a way to control the competition of Russian oil in world markets.

Efforts from 1928 to 1930 were directed primarily toward controlling world production and supplies from all three of these sources. Private agreements to limit production were negotiated with outsiders in some foreign countries. (The Roumanian agreements of 1930 to 1932 are examples.) Russian competition, however, still continued to be troublesome, especially in European markets. In the United States, lawful ways of controlling production and exports were sought. The American Petroleum Institute sponsored proration of production in the late 1920's and sought both State and Federal sanction for its scheme as a conservation measure. This movement later developed into the system of interstate oil compacts sanctioned by the Congress as a conservation measure in the national interest. A few years later, the international companies similarly urged worldwide conservation as a justification for private agreements to limit production in foreign countries. Thus conservation became the cartel's slogan at a time when a rising flood of international production threatened to depress world prices.

In the United States, the conservation movement was supplemented in 1928 by the formation of Export Petroleum Association, Inc., as a lawfully constituted Webb Act association. This organization was to control American exports and fix the f.o.b. Gulf prices, which the principal international companies had agreed should be the base prices used in determining the value of oil products of uniform specification in all world markets. When Export Petroleum Association's members were unable to agree upon prices among themselves, due to foreign interference, and the legality of the proposed price fixing activities was questioned, the Association's activities collapsed in November 1930. Thereupon, the 1930 Roumanian Agreement, the success of which depended upon the enhancement and maintenance of American export prices, also came to naught.

The experiences of the three principal international companies during the first two years of operation under the Achnacarry Agreement were such as to make it clear that their control over the major sources of production outside the United States and Russia was, by itself, insufficient to serve as a basis for world-wide stabilization of the market. They could not always agree among themselves and there was an increasingly important uncontrolled fringe of producers and marketers of various nationalities who were not bound by the main agreement. The activities of outside producers seldom were world-wide, but by cutting even slightly under the cartel's prices in those markets in which they were interested, the outsiders became the beneficiaries of the cartel's program. If the cartel sought to drive such unwelcomed outside competition out of one market by price competition, the independents had the tactical advantage of being able to retaliate in other markets in which they previously had not been factors. The effort to stabilize the industry through a single world-wide agreement, therefore, demonstrated the difficulty of aligning diverse interests under a single agreement, and the insufficiency of the dominant position of three companies in world production and distribution to serve as the basis for effective cartel controls in local markets.

The Big Three conferred late in 1929 in an effort to implement the principles of their 1928 agreement. The result was the Memorandum for European Markets of January 20, 1930, in which the three major international companies set out the terms upon which they agreed to act as a unit in attacking the problem of world control (a) through agreements among themselves and with others to control production in particular producing areas and (b) through marketing agreements involving themselves and others in particular consuming countries.

The first attempt under the expanded plan was the formation of a new Roumanian Agreement in 1930, under which all Roumanian producers agreed to limit production and fix prices for oil from the Roumanian field. The success of this agreement hinged upon finding markets for Roumania's controlled production largely in European countries at prices satisfactory to Roumanian producers. As already indicated, the Roumanian Agreement fell when Export Petroleum Association, Inc., was unable to agree on f.o.b. Gulf prices and became inactive. Thereupon, Steaua Romana, an affiliate of Anglo-Iranian, withdrew from the Roumanian proration scheme.

Concerted effort to bring Russian and Roumanian independent production under control and find a market for it at prices fixed by local marketing cartels formed in accordance with the 1930 Memorandum for European Markets, however, did not cease. Partial cooperation by the Russian trading company was obtained in some markets, and a new Roumanian production agreement was made in 1932, which the Roumanian independents signed only on condition that they be given a larger production quota than they would have been entitled to with the year 1928 as the base. This constituted a departure from the "as is" procedure previously laid out in the Achnacarry Agreement. Furthermore, during 1930 to 1932 local marketing agreements to which outsiders were admitted proved difficult to conform to "as is" as set out in the Achnacarry Agreement and the Memorandum for European Markets. This was so because of numerous concrete problems of procedure in sharing markets, fixing prices, making financial adjustments for under- and over-trading and adjusting quotas in markets where a participant entered a local market in which he had not previously sold or where a participant acquired a previously existing independent outlet in a local market in which he already was selling.

The development of rules to solve these problems was attempted in a new international agreement known as Heads of Agreement for Distribution, which was adopted on December 15, 1932, at a meeting attended by representatives of Socony-Vacuum, Standard Oil Co. (N.J.), Anglo-Persian, Shell, Gulf, Atlantic, and Texas. The new agreement provided greater elasticity in the application of the Achnacarry Agreement's quota provisions by setting out in detail the manner in which the participants to the main agreement would contribute from their quotas in local markets to make room for larger production quotas granted to outsiders, as had been done in the case of Roumanian producers. Other revisions and additions made it clear that the revised international understanding covered the following points:

(1) That the principle of "as is" was to apply to every country or area of the world, except the United States.

(2) That it was to apply to supply (i.e., production and exports) and distribution of crude oil and refined products.

(3) That supply "as is" would be handled by a committee sitting in New York.

(4) That distribution "as is" would be handled by a companion committee sitting in London.

(5) That the activities of these two working committees would be coordinated through a central "as is" committee sitting in London.

(6) That all disputes arising under the Heads of Agreement would be decided by a central "as is" committee which would also provide rules for meeting problems arising as a result of new factors or changing conditions in the industry.

Thus machinery such as had not previously existed was provided for coordinating world supply agreements among the major international companies with local marketing agreements to which outsiders were also to be admitted.

During the next year and a half, comparatively little was accomplished in further cartelization under the Heads of Agreement because, with American production uncontrolled, f.o.b. Gulf prices, which, by international agreement, were the base prices for computing European market prices, fell as low as 30 cents per barrel in May 1933. The disguised independent Roumanian producers thereupon refused to observe their agreement to limit production, and shipped a few cargoes to markets as distant as Canada and Australia. Resistance to local marketing agreements based on "as is" also arose in European consuming markets.

Shortly thereafter, however, U.S. Gulf prices were advanced under the NRA Code of Fair Competition from 50 cents in July to $1.00 per barrel in

October 1933, and became stabilized at that level. Thereupon, in February 1934, it was reported that the Roumanian producers were as willing as ever to control their production provided foreign production and prices were brought under control.

In the meantime, the three major international groups had tightened their joint control over reserves and production in Venezuela and the Middle East and continued their dominant position in Roumania and the Far East. The development of local market agreements became feasible again.

Past experience had demonstrated that although the "as is" principles might be strictly controlling with respect to relations among the Big Three, the diversity of interest of independents with whom these companies necessarily had to deal in local markets was such as to necessitate regarding those principles as ideals to be attained as far as practicable, but not to be insisted upon too strongly in dealing with outsiders. For this reason, as well as to clarify the positions of the principles with respect to each other, a new agreement known as the Draft Memorandum of Principles was negotiated in June 1934.

This agreement, the fourth, longest, and last of the prewar international understandings, was essentially a restatement of the "as is" principles of previous agreements. New Material dealt largely with procedural amendments and detailed rules for the application of the principles in all markets of the world except the United States and any other countries where their application by agreement would be unlawful. The new rules dealt largely with special conditions under which the application of "as is" had proved troublesome in the past. Specifically, it was agreed that:

(1) Local agreements to which outsiders were admitted were to be regarded as wholly separate from the Draft Memorandum, and outsiders were to have direct knowledge of the main agreement only to such extent as the London executives of the major parties might direct.

(2) In any market where all parties to the main agreement already were, or might subsequently become members of a local cartel including outsiders, the major parties might unanimously agree that the local cartel would override the memorandum for that market for the duration of the local agreement.

(3) Two types of quotas for the participants were to be set up in each local market, one covering the total quantity supplied by each participant, including sales to other participants and approved outsiders, and the other covering only quantities actually distributed by each participant.

(4) Fines assessed against excessive over-traders were to be apportioned back to under-traders on a more equitable basis than had been provided under previous agreements.

(5) Quotas were to be adjusted for each of the following reasons:
    (a) As a penalty against under-traders the quantities required to keep the quotas of all participants in balance:
    (b) When new members were admitted to the local agreements,
    (c) When a participant acquired an outside outlet by purchase
    (d) When a participant voluntarily retired from the market;
    (e) When one product replaced another in use in the market (for instance, if Diesel fuel oil replaced gasoline).

(6) Under certain conditions specifically described, the general rule that price movements, either up or down, were to be governed by a majority vote of all participants in each local market was to be suspended, but even at such times, the practice of open price reporting and discussion of proposed prices in advance was to be followed.

(7) After a specified date no competitive expenditures for advertising and sales promotion and for capital investment for facilities were to be made except in accordance with budgets previously approved by the cartel's London committee.

This partial review of the various international agreements indicates that from the comparatively simple proposal of the Achnacarry Agreement in 1928 to stabilize the world market by controlling production and exports there was evolved in 1934 an international agreement which contemplated detailed restriction of production, division of markets, price fixing, restriction on the number and kind of distribution outlets, and the elimination of competitive expenditures for market facilities and sales promotion in local markets.

So far as the record shows, the three prime movers in the formation of the international agreements, from Archnacarry in 1928 through the Draft Memorandum of Principles in 1934, were the three major international companies. This appears to have been true even though some other important American companies sat in the conferences which formulated the later international understandings having to do with local markets. Jersey Standard, which appears to have acted as the leader among the American companies, states that it gave verbal notice of its withdrawal from the Draft Memorandum agreement in 1938, and that any activities which may have survived came to an end at the outbreak of the war in September 1939. From this, it might be inferred that cartel cooperation came to an end at the outbreak of the war.

However, British and Dutch cooperative relations were by no means severed with Standard's withdrawal, and American companies continued to cooperate to some extent both during and after the war. In the United Kingdom an industry committee, acting with the approval of His Majesty's Government, set up and operated a wartime pooling plan for international supplies and local distribution modeled quite closely along "as is" principles, and in Sweden, a neutral country, it was not until 1942 that American companies informed their subsidiaries that local agreements of the type made under the Draft Memorandum must no longer be made for the reason that they might be in violation of the American antitrust laws. Even then, cooperation did not entirely cease, for a Swedish investigating committee found that adjustment of under- and over-trading in accordance with "as is" was considered in 1943, and that in 1946 prices and terms of sale for the year 1947 were agreed upon.

More than 10 years of cumulative effort to operate and extend controls under the various international agreements apparently established acceptance of the durable nature of "as is" principles, so that their observance among the corporately interrelated international group became practically a custom of the trade. The development of the cooperative atmosphere was favored by the intricate [maze] of contractual and other relations developed over the years among the small number of large international oil interests.

Chapter IX—Case Studies in the Application of Marketing Agreements in Selected Areas

The structure of control and agreement in the international petroleum market was made complete during the inter-war period by the organization of local marketing cartels, based upon "as is" principles, in most countries of the world. The elements of this structure have been shown in the previous chapters of this Report as including:

(a) The predominant and strategic position of seven international oil companies in the ownership of world petroleum reserves, production, refining capacity, and transportation facilities.

(b) Joint ownership by these companies in affiliated and subsidiary companies in all parts of the world, particularly in the Middle East, and in all phases of the industry.

(c) Crude oil sharing arrangements in important producing regions.

(d) The development of agreement upon a philosophy and strategy of cooperation in marketing the end products of the industry. The local marketing cartels described in this chapter flowed naturally from these developments and provided an additional and important element in the structure of control and agreement.

This chapter, therefore, deals with: (1) an examination of the procedure followed by the subsidiaries of the international oil companies in creating and maintaining local marketing cartels in accordance with the principles set forth in the international marketing agreements described in Chapter VIII, and (2) an examination of the scope and effectiveness of the local cartels.

The marketing arrangements described in Chapter VIII were, for the most part, general statements of policies and goals. They were designed to set up a broad and flexible framework of principles to be applied in each of the particular markets in which these oil companies were interested. The Achnacarry ("as is") Agreement, the Memorandum for European Markets, and the Draft Memorandum of Principles, therefore did not directly allot quotas or fix the conditions of sale, but, rather, constituted guiding statements to be used in preparing and administering *local arrangements* to be put into effect for each petroleum product in each marketing area.

The international marketing agreements further provided that where the participants in the agreements were or intended to become members of local cartels, the local cartel arrangements would "override" the "as is" principles by unanimous consent. Thus the international agreements were highly flexible documents and the intention was to apply the agreed marketing principles, subject to special modifications or substitutions in the appropriate cases to all countries in which the international oil companies had marketing interests, except the United States.

The application of the broad agreements discussed in Chapter VIII was effectuated only after a lengthy and continuous process of negotiation and consultation by local managements in each marketing area. It was necessary to agree upon marketing arrangements for each country, guided by the broad and flexible directives given in the international agreements. These local arrangements were to be governed, administered, and modified in the light

212 Appendix

of current conditions by the local officers of the international oil companies, who were directed to meet frequently for that purpose.

The lengthy process of preparing and administering the local arrangements is illustrated by the case of Sweden. Fairly complete information is available with relation to that country as a result of an investigation and report by a legislative committee in 1947. Sufficient information exists about other countries to indicate that the Swedish case history resembles that of other local cartels that operated under the terms of the international agreements in other foreign countries.

The "case studies" made in this Chapter involve market arrangements or local cartel operations involving Sweden, United Kingdom, France, Germany, Belgium and the Netherlands, Scandinavian Countries (Denmark, Norway and Finland), Argentina, Chile, Brazil, Mexico, Cuba, the Lesser Antilles and other countries in Europe and Asia.

The principal case study involves Sweden which was the subject of investigation by The Oil Investigating Committee of 1945. This committee, the functions of which were to investigate oil enterprises in Sweden, was appointed pursuant to authorization of the Rikstag of Sweden in 1945.

The summary and review of the Swedish case history is as follows:

The Swedish case history has demonstrated in some detail the processes by which the local marketing subsidiaries of the international oil companies have negotiated and put into effect cartel arrangements. The major conclusions to be drawn from this case history are as follows:

(1) The oil companies participating in the cartel arrangements effectively dominated all petroleum product markets in Sweden from 1933 to 1936. In each market, except asphalt, the cartel participants, including Gulf, delivered in 1936 from 96 to 100 percent of the tonnage sold for domestic consumption in Sweden.

(2) The Swedish subsidiaries of Standard (N.J.), Shell, and Anglo-Iranian—the "as is" group—participated in all of the reported negotiations, agreements, meetings, and cooperative efforts of the oil companies in Sweden during the period under review in this case history, 1930–1947. Texas was an equal participant from the time it became a major factor in the Swedish market in 1937. Nynas, a Swedish concern, and Nafta, the Russian owned predecessor of Gulf, collaborated in the main with these firms. Gulf's collaboration with the cartel began at the latest in January 1936.

(3) The Swedish cartel arrangements were inspired and guided for the most part by the international "as is" agreements, particularly the Draft Memorandum of Principles.

(4) The cancellation of the Draft Memorandum of Principles at the end of 1937 was not intended to, and, in fact, did not, result in the termination of the local agreements, since the parent companies instructed Standard (N.J.), Shell, and BP to continue the local agreements, if possible. The five-party Swedish "as is" agreement continued until at least the end of 1938. Repeated efforts to draw up a new five-party or, failing this three-party agreement were interrupted by the outbreak of World War II. The oil companies prepared a new price agreement after the war, which was intended to be effective on January 1, 1947.

(5) The parent corporations supervised and directed the cooperative efforts of their subsidiaries in Sweden. While the Oil Investigating Committee was able to investigate this matter in only one case, they report that "the

company concerned has, with regard to cooperation with other companies, in some cases received detailed instructions from their principals, and has in turn rendered detailed reports to them and asked for their approval of what has occurred.

(6) The administration of the Swedish cartel arrangements was carried on through the medium of regular weekly meetings of the managing directors of the companies, at which all matters of common and current interest were discussed and differences of interest or interpretation were resolved through negotiations. The Investigating Committee likened these meetings to "a permanently functioning board of cooperation with Shell as the executive organ." At the meetings, the companies regularly rendered audited accounts of their trade statistics, i.e., "lists of deliveries invoiced" by them.

(7) The cartel arrangements included agreements that the companies would charge identical prices and would fix uniform rebates, commissions, bonuses, discounts, and other selling terms. It was agreed that customers would be classified and that rebates, and the like, would be eliminated for some classes of special customers and made uniform for others. It was agreed that the companies would consult in advance before any firm made tenders to State institutions and other large customers. It was also agreed that the companies would "respect" each other's customers.

(8) The distinctive "as is" arrangements applied at first only to the "as is" group, but in 1937 were extended to include Texas and Gulf. Nynas did not adhere to these arrangements but cooperated in other respects. The adherents to the "as is" arrangements were assigned distribution quotas in each product market. A system of adjustment and compensation for over- and under-trading was instituted. The principal method of adjustment agreed upon was the transfer of major customers, particularly State institutions and the like.

(9) The Investigating Committee found that the arrangements set forth in paragraphs (7) and (8) above were generally adhered to by the oil companies. Price changes in the market were the result of decisions of the companies. Uniformity of prices, rebates, and sales conditions was largely achieved. Special customers were classified and their terms of sale fixed as a result of negotiations among the oil companies. Markets were divided and customers steered to the designated suppliers by a number of devices, including that of tenders at fictitious prices. Adjustments of over- and under-trading were made, often by the device of transferring customers. While the companies contended that the agreements were, in effect, mere paper promises, the committee was satisfied that they had been substantially put into effect.

(10) Outside marketers, such as the consumer cooperatives and the small importers, were generally whipped into line in order to protect the interests of the oil companies, including the cartel arrangements, and the interests of the affiliated outlets. The committee reports that "decided pressure" was exerted on these outsiders to adhere to the prices and other matters in the agreements. The companies refused to do business with the cooperatives, except the IC, or proposed terms which were onerous, or required the cooperatives to limit their activities. Special efforts were made to bring the IC into cooperative agreement with the oil companies. In nearly all cases control over supplies was the chief weapon used.

(11) Since the Investigating Committee was unable to obtain essential

data relating to costs, it was unable to state whether prices had been unreasonable or the oil enterprises unduly profitable. According to a statement by one of the companies, benzine prices in 1939 were unduly high but no reduction in price was made. State and municipal institutions were especially vulnerable to the cartel operations, but the effects on consumers in general could not be determined.

The above summary recapitulates conclusions unanimously arrived at by the Oil Investigating Committee of 1945. The Swedish oil market was not a free market but one regulated and controlled by private arrangements. The primary task of the committee was to survey the oil trade and to submit proposals with regard to the future of that business; among the alternatives considered was that of a State monopoly of the trade. These facts were among the important considerations that led the majority of the committee to recommend the establishment of a State monopoly.

The conclusions drawn from all of the case studies contained in this chapter are as follows:

The case studies given in this chapter illustrate the wide-spread development of local cartel arrangements following the conclusion of the Achnacarry or "as is" Agreement of 1928. As has been noted, this international agreement on marketing principles was inspired by the price war which had broken out in the Far East, particularly in India. Its principles, which were elaborated upon and modified in subsequent international agreements in 1930, 1932, and 1934, proved applicable in all kinds of markets, in large countries and small, in industrialized economies and in agricultural and even "undeveloped" economies.

There can be no doubt that most of the local cartel arrangements were guided by the international agreements. In nearly all petroleum markets of the world, outside of the United States, subsidiaries and affiliates of the principal parties to the international agreements—Standard (N.J.), Royal Dutch-Shell, and Anglo-Iranian—were predominant forces. Cartel agreements would have been unworkable without their leadership and cooperation. These companies were closely associated in their direction of marketing activities from London—the three companies being described by a Jersey Standard official as a "joint venture"—and were bound to apply the "as is" principles in all local markets. The cartel arrangements universally included quota arrangements designed to freeze the market pattern—the historical position of the marketers—in accordance with that of a "base year" or "qualifying period," which in most cases was the year prescribed in the international agreements, 1928. Most of these distribution quotas were subject to a general revision, effective January 1, 1936, although nearly all changes were minor in character. Other features of the cartel arrangements included provisions to protect the division of the market, such as the agreements to "respect" each other's customers, the fixing of prices, schedules of rebates, discounts, etc., and other selling conditions, the adjustment of quotas for over- and under-trading, the application of a system of fines and compensation in adjustment of over- and under-trading, and the limitation of competitive expenditures for marketing facilities.

In most cases where there were substantial "outsiders" in the market, they were brought into the cartel. These important outsiders were occasionally successful in causing modifications or substitutions in various "as is" provisions, but the central principle, that of preserving the historical positions of the participants, always remained. These modifications and substitutions were expressly permitted by the international marketing principle that the interested "as is" parties could enter into such local cartel arrangements by unanimous consent.

This analysis of the origin of the local marketing cartels of the 1930s does not accord with the interpretation advanced by the international oil companies themselves. Lawrence B. Levi, a director of Socony-Vacuum and its executive in charge of foreign operations, speaking for a committee of American oil companies stated in 1945 that the cartels developed out of the following factors: the development of excess productive capacity after the first World War, the relative narrowness and leanness of most foreign markets in comparison with the United States, were price cutting and "dumping" by the Russians, causing "price wars and wasteful practices," and the development of national economic policies including such matters as currency and monetary controls, and regulation of markets. He did not mention international marketing agreements, but rather laid great emphasis on the role of governments desirous of attempting "market stabilization within their countries."

Mr. Levi then went on to state:

"Trade agreements which affected the oil industry in foreign countries may be classified under two general headings:

"1. By law, i.e., Government monopoly or compulsory cartel.

"2. (a) Trade agreements by Government direction or pressure, and (b) Trade agreements organized by private initiative in accordance with the law of a country and permitted or encouraged by its Government."

Thus he interpreted the local cartels as merely the result of compulsive governmental or social forces within each local marketing area. American companies he said, were forced to go along with these restrictive forces if they wished to stay in business in these countries.

This interpretation omits the role of the international oil companies themselves in the development of the local cartels, as described in Chapters VIII and IX of this report. The international agreements were developed partly as instruments to deal with the problems listed by the industry spokesman. These forces were conditioning and limiting factors that determined the form of application and the effectiveness in each case of the cartel arrangements under "as is." It would be difficult indeed to list many of the cartels described in this chapter under the categories listed by the industry spokesman.

The assertion that local customs and usages determined the growth and form of the marketing cartels must also be questioned. In this regard the significance, effect, and form of the local marketing arrangements of the 1930s is much the same as that of the international "as is" agreements. The international "as is" agreements largely determined the character of local cartelization. Local cartel arrangements pursuant to the "as is" agreements were themselves largely determinative of the customs of the trade.

Chapter X—Price Determination in the International Petroleum Industry

The preceding chapters have described the degree of control over the world petroleum industry held by the seven major international oil companies, their participation in joint-ownership ventures, as in the Middle East, their control over sources of supply through contracts among themselves for the purchase and sale of crude oil, as in the Middle East and Venezuela, their development of world-wide production and marketing agreements, and the application of these agreements in specific countries. Moreover, it has been shown that these major oil companies, through the high degree of concentration of control, through direct ownership, through purchase and sales contracts, through joint ownership, and through production and marketing agreements, have been able to limit production, divide up markets, share territories and carry on other activities designed to stabilize markets and control production.

In addition, the international petroleum companies have followed a system of pricing which has had the effect of eliminating price differences among themselves to any buyer at any given destination point. Under this system the delivered price to any given buyer is exactly the same regardless of whether he purchases from a nearby-low-cost source or from a distant high-cost source. While the delivered price from all sellers will be different at one destination point as compared with another, reflecting largely differences in freight costs, their delivered price at any *given* destination point will be exactly the same.

This systematic elimination of price differences has been achieved through the establishment and observance of an international basing point system, generally referred to as "Gulf-plus." In recent years the system has undergone a number of minor modifications. Nonetheless, since each of the major companies has usually observed these modifications, the system's ultimate effect upon any given buyer has remained the same—the elimination of price differences as among the various sellers.

Under basing systems the quotation by sellers, no matter where located, of identical delivered prices at any given point of destination is arrived at in the following manner:

A particular producing center (or centers) is designated as "the base point," at which a "base price" is established. The various sellers then arrive at a delivered price by adding to the "base price" the freight charges therefrom to the point of destination. Those sellers who are located nearer to the buyer than the "basing point," reap the advantage of "phantom freight"—i.e., the difference between their actual freight charges to the buyer and the freight charges from the "basing point," to the buyer. Similarly, those sellers who are located farther from the buyer than the "basing point" have to "absorb" freight—i.e., they have to absorb the difference between their actual freight charges to the buyer and the freight charges from the "basing point" to the buyer.

As can thus be seen, the starting point in the operation of a basing point system is the establishment of the "base price" at the "basing point." Hence, the first step in discussing a basing point system should be an examination of the adequacy and representativeness of this price. In the international petroleum industry the "base point" has generally been U.S. Gulf, and the "base price" established at U.S. Gulf has been derived from an American trade Journal.

The trade journal from which the "base prices" are derived in the petroleum industry is *Platt's Oilgram Price Service,* published at Cleveland, Ohio. It has been appropriately described by an industry source as "the framework on which this complex international price structure is carried." The same source further described Platt's price reporting service as follows:

Appendix 217

"This is a daily publication of oil prices in the United States, both for the home market, where quotations are given at a number of important sources of supply and centers of consumption, and for the export market where prices are quoted f.o.b. the major oil ports . . . .

"Most long-term sales contracts of petroleum products are linked throughout the world, to Platt's f.o.b. prices, and the price fluctuates directly in proportion to the changes in Platt's.

"*Even where, for any special considerations, the price of a product is not the same price that it would have been if it had been shipped from the United States,* allowing for freight charges, it is frequently the practice, in a contract which is to spread over a period, to link *the price to the Platt's prices* of a grade selected for reference purposes. This means that even if the sale is not originally directly related to the United States price . . . it is usual in international trade to allow subsequent fluctuations of that price directly in proportion to the movements in Platt's." ("The Price Structure of the Oil Industry," by R. C. Porten, published in *Oil*, house organ of the Manchester Oil Refinery, Ltd., England.)

A summary of the matters respecting pricing petroleum and its products which are treated in this chapter is as follows:

The use of the Gulf-plus basing point system, both in its original and modified forms, to price crude oil and refined petroleum products served two basic purposes of the major oil companies:

(1) It eliminated differences in delivered prices among the various sellers at any given point of destination, thereby making the selection on one seller over another a matter of indifference to the buyer insofar as price was concerned.

(2) It made the relatively high United States Gulf prices the basis for both crude oil and refined products prices through the world.

The first break in single basing point pricing occurred during World War II when the Persian Gulf was made a basing point with prices equal to those prevailing at United States Gulf ports. In 1943, when the British Government began buying large quantities of bunker fuel in the Persian Gulf for the use of its Navy, British officials objected to the large amounts of phantom freight involved in buying these products on the basis of Gulf-plus and insisted that the Persian Gulf be made a basing point. The suppliers thereupon established the Persian Gulf as a basing point with base prices for refined products equal to those quoted by *Platt's Oilgram* for United States Gulf ports. Later in 1945, on sales to the U.S. Navy, American companies operating in the Middle-East likewise established a Persian Gulf base price for crude oil and refined products equal to the United States Gulf price.

Under the dual basing point system thus set up, the point of equalization, i.e., the point in the important European consuming market at which Persian Gulf products and United States Gulf products were delivered at equal prices, using USMC transportation charges from both basing points, was the mid-Mediterranean—i.e., shipments from either producing area beyond that point required the absorption of freight. Postwar reopening of western European markets which had been closed to Middle East producers during the war made this absorption of freight important especially to Middle East producers when they sought to supply European markets from their rising production of Middle East crude and refined products.

Between November 1946, when OPA price controls were terminated, and April 1948, United States Gulf prices for crude oil and refined products moved sharply upward. Persian Gulf base prices for refined products continued to be equal to those quoted in *Platt's Oilgram*, "Platt's high" being used as the Persian Gulf base for shipments to China, "Platt's mean" for shipments to other destinations east of Suez, and "Platt's low" for destinations west of Suez, i.e. mainly the Mediterranean and Europe. Thus, the Persian Gulf base prices for refined products remained equal to United States Gulf prices.

Advance in the Persian Gulf base price for crude, however, lagged behind the advances in U.S. Gulf crude prices, and finally became stabilized at about the end of 1947 at $2.22 per barrel. This price, although lower, was tied to the United States Gulf price by a definite pricing formula under which the price of crude originating in both areas was equalized at the United Kingdom, thereby opening a wider market for Persian Gulf oil without absorbing freight under the basing point formula. The formula consisted of adding to the quoted base price for Venezuelan crude the USMC freight rate to the United Kingdom, and deducting therefrom the USMC freight from the Persian Gulf. This yielded the base price for crude at the Persian Gulf. Thereafter, this formula, sometimes using USMC rates and sometimes using lower freight charges when open market charter rates fell below USMC rates, was systematically used to link the Persian Gulf base price for crude to the higher base price prevailing at United States Gulf ports. This same formula was used in July 1949 when, in recognition of the fact that Persian Gulf crude was moving to the United States in quantity, the point of equalization was moved to New York. This established a base price of $1.75 per barrel f.o.b. the Persian Gulf (Ras Tanura) which thereafter remained unchanged up to the time this report was prepared (August 1951).

When open market charter rates advanced sharply after 1949 application of the formula would have reduced the Middle East base price by about 85 cents per barrel. The formula, however, was not applied and the Persian Gulf base price remained unchanged. Moreover, when the Trans-Arabian Pipeline from the Persian Gulf to Sidon on the Mediterranean was completed in December 1950, the f.o.b. price $2.41 established at Sidon was based on the $1.75 Persian Gulf price plus water freight from Ras Tanura to the eastern Mediterranean, including the Suez Canal toll. In other words, the tanker transportation charge was made the rate for transmitting crude oil by pipeline. This avoided establishing different prices in the eastern Mediterranean for oil transported by pipeline and by tanker, but it also enabled the companies owning and using the pipeline to keep for themselves all savings of pipeline transportation over tanker freight. Also, for more than a year after the $1.75 price for crude was established at the Persian Gulf, Middle East refined products continued to be quoted at *Platt's Oilgram* prices plus whatever freight charges were currently being used to determine delivered prices.

When ECA began financing shipments of refined products it found that the use of straight basing point pricing resulted in price discrimination as between countries to which it financed shipments. To correct this, it divided the European market into zones and ruled that the price it would pay for Persian Gulf products delivered in a given zone would be no more than the realized net-back for Persian Gulf products computed by deducting the USMC rate from the Persian Gulf to the zone from the alternative delivered price for Western Hemisphere oil to that zone. Since USMC rates were higher than either company rates or open market charter rates, this reduced the net realization of Middle East suppliers of refined products financed by ECA. The industry objected strenuously to this reduction.

Throughout the period under review, the major international oil companies have clung tenaciously to the basing point method of pricing. Regardless of the fact that they modified the single basing point system and made other concessions largely as the result of government pressures, the existing price structure is still highly profitable to the small number of major international oil companies that dominate world production. Under the resulting price structure, American companies operating in the Middle East have made substantial net profits on their combined producing, refining and marketing operations. Thus, according to the published statements of Standard Oil Co. of California and The Texas Co. it appears that these and other joint owners of Aramco and Bahrein Oil Co., Ltd., realized net profits amounting to about 91 cents per barrel of crude oil produced, refined, and marketed by these two companies in 1948; 95 cents in 1949, and 85 cents in 1950.

These profits were realized under a system of pricing that:

(1) Bases delivered prices throughout the world on the relatively high U.S. costs, notwithstanding the fact that this country has become a net importer of petroleum.

(2) Uses schedules of uniform freight charges that may not have any real relationship to transportation costs actually incurred, especially by the major companies that own or control the bulk of the world's tanker facilities.

(3) Is supported and maintained by effort on the part of the major international companies to adjust production to world demand. Joint ownership and private agreements in foreign countries and the conservation movement in the United States all facilitate these efforts.

Thus, although a new basing point was established, although the point of equalization was changed on several occasions, and although other modifications were made, such as the use of arbitrary percentages of USMC freight rates in computing delivered prices, the essential character of the "Gulf-plus" basing-point system has remained unchanged. It still fulfills its basic purposes of eliminating price differences among sellers to any given buyer, and of making the U.S. Gulf price the principal determinant of the *level* of world prices. As such, it is highly profitable to the international companies. By performing these functions, the system serves as a highly useful complement to the other types of controls exercised over the international petroleum industry by its major companies.

The propriety of continuing into the future this pricing system has been questioned by industry spokesmen themselves. Thus in describing the underlying reasons for deviations from a strict use of "Gulf-plus," Dr. Frankel in 1948 stated:

"During the last two years or so, factors have come up which, severally and jointly, have begun to render the erstwhile set-up more and more obsolete: firstly, the U.S.A. has become a net importer of petroleum, which makes it more difficult to maintain the conception that the U.S. Gulf is, in fact, the fountain head of the world's oil; secondly, moving inversely, the Middle East crude output has risen to such an extent that it is bound soon to cover the Eastern Hemisphere demand and can no longer be considered to be

220   Appendix

supplementary in its scope; thirdly, the repeated increases in the domestic American price level, which took place in 1947, and which were determined by domestic causes, and which may be followed by further similar moves, have greatly widened the gap between similarly increased Middle East (and for that matter, Latin American) crude prices and the level of cost of production as it is known or as it has been estimated in the past."

After considering somewhat sympathetically the rationalization by which Middle East producers still adhere to this system, Dr. Frankel called attention to the basic weakness of the argument by pointing out that:

"Such conceptions, however rational they may be from the point of view of the operators themselves, do not entirely meet the case when it comes *to facing the responsibilities of big companies toward consuming interests.*"

WORLD CRUDE OIL PRODUCTION BY AREAS, 1946–67
(Million barrels per year)

Year	Middle East	North America	South America	North Africa	All other areas	World total
1946	252	1,790	466	9	234	2,751
1947	304	1,920	519	9	252	3,004
1948	417	2,091	556	13	336	3,413
1949	511	1,924	574	16	374	3,399
1950	636	2,073	638	16	420	3,783
1951	702	2,369	726	17	463	4,277
1952	763	2,428	766	18	521	4,496
1953	885	2,510	756	18	601	4,770
1954	999	2,495	810	15	712	5,031
1955	1,180	2,702	910	14	836	5,642
1956	1,263	2,881	1,036	13	908	6,101
1957	1,294	2,891	1,169	17	1,157	6,528
1958	1,559	2,715	1,121	26	1,182	6,603
1959	1,679	2,856	1,208	31	1,336	7,110
1960	1,924	2,864	1,265	90	1,513	7,656
1961	2,054	2,950	1,319	155	1,668	8,146
1962	2,258	3,032	1,439	260	1,863	8,852
1963	2,486	3,125	1,468	393	2,005	9,477
1964	2,782	3,177	1,531	564	2,210	10,264
1965	3,051	3,259	1,565	692	2,443	11,010
1966	3,409	3,470	1,549	856	2,678	11,962
1967	3,658	3,700	1,649	997	2,873	12,877

Source: *"World Oil" Atlas,* 1947 and 1948, International Outlook issues, 1950–68.

Index

Acheson, Dean: xii, 34, 35
Advertising: 61, 123, 125, 126, 151
Alabama Power Co.: 55, 59
Alaska: xii, 72
Alternatives: energy sources, 115–119, 159, 161; energy system, 154–174; planning, 170
Amalgamated Copper Co.: 5
American Electric Power Co.: 53, 166
American Gas Association: 86, 87
American Gas and Electric Co.: 57
American Petroleum Institute: 7, 9, 20
American Power and Light Co.: 57
American Super Power Co.: 57
Andrews, Samuel: 2
Anglo-Iranian Oil Co.: 15, 26, 27, 30, 31, 34, 35
Anglo-Persian Oil Co.: xi, 15–16, 20
Antitrust Division (Justice Dept.): xii, 23, 30, 34–38, 81
Arab boycott: 1, 90, 91
Aramco: 1, 21, 24–26, 27, 31, 32
Atomic Energy Commission, energy needs: 64, 79, 166

Banks, interests in power: 4–5, 65, 66–67, 75, 129, 131
Berkeley, resistance in: 147–148
Bidding combines: 73–74, 75
British Development Corp.: 21
Brugman, Bruce: 146
Burlington Northern Railroad: 83
Burmah Oil Co.: 15

Canada: xii, 72
Carey, Edward M.: 96, 97
Cartels: 15–22, 26, 75–78, 94–98, 111, 114; opposed, xii, 34–38; origins, xii, 1, 13, 15–19
Charter, Boyd and Ann: 84–85
Chesapeake region, dependency of: 154–161
Church, Frank: 1, 33, 94
Churchill, Winston: 15–16
Cities Service Co.: 49, 58, 93
Clark, Tom: 28
Clarke, M. B.: 2

Coal gasification: 39, 83, 85–88
Coal industry: 39–46, 78–81; boom, 81–85; government and, 45–46; surface leasing, 82–85
Colorado Interstate Gas Co.: 48, 49
Columbia Gas and Electric Corp.: 49
Commoner, Barry: 157
Commonwealth Edison Co.: 55
Competition: coal, 40, 41, 45; industry aversion to, 27, 71, 76
Concessions Syndicate, Ltd.: 15
Connally Act: 10
Conservation: 7, 10, 44, 81, 103, 106
Consolidated Gas Co.: 5
Consolidation Coal Co.: 45, 80, 84–85
Consumer Energy Act: 109, 111
Continental Oil Corp.: 45, 80
Coolidge, Calvin: xi, 9
Corey, Kenneth: 75

Daniels, Josephus: 8
D'Arcy, William Knox: 15
DeGolyer, E.: 7
Delaware and Hudson Canal Co.: 40
Dellums, Ron: 141
Depletion allowance: 9, 113–114
DeYoung, Joseph Y.: 146
Dixon-Yates Contract: 64–65
Doheny, E. L.: 9
Doherty-Morgan-Ryan Co.: 55
Drilling costs, deduction: 114
Dulles, Allen: 37

Eaton, Cyrus: 56
Eavenson, Howard: 41–42
Edgar, E. MacKay: 18–19
Edison Illuminating Co.: 5
Eggers, Alfred J., Jr.: 116
Eisenhower, Dwight D.: 36, 51
Electric Bond and Share Co.: 53, 57, 58, 60, 65, 66
Electric Power and Light Co.: 57, 58
Electric utilities: 52–67; public, 52, 61–63, 139; transmission, 52, 62, 83
El Paso Natural Gas Co.: 88

221

222 Index

Energy crisis: 89–100; contrived aspects, xii, 1, 91–94, 98–100; warning signs, xii, 18, 86
Engle, Clair: 145
Environmental Action Foundation: 150–151
Environmental Policy Center: 103, 105, 152
Environment Magazine: 98–100
Ethyl Gasoline Corp.: 13–14

Fall, Albert: 8–9
Federal Energy Office: 107, 109
Federal Oil Conservation Board: 9
Federal Power Commission: 50, 55, 56–57, 85, 87–88, 168
Federal Trade Commission: 34, 57, 129
Flagler, Henry M.: 2
Flanigan, Peter: 92
Ford, Gerald R.: 106*n*
Foreign tax credits: 32–33, 114–115
France: 17, 21–22
Freeman, David: 110, 140
Fuel adjustment clauses: 96, 129, 139, 150
Fulbright, William: 50

Gasoline: 12, 13–14
General Electric Co.: 53, 55
Georgia, resistance in: xiii, 122–125
Georgia Power Project: 122–135, 160
Government, federal (*see also* specific units), as industry ally: xi, xii, 23–24, 26–27, 32, 72–74, 82–88, 98, 171
Government, state: ally of industry, xi, 6–7, 171; prorationing laws, 10, 20, 90; tidelands oil issue, 28–29
Great Britain: xi, 15–17, 45
Gulbenkian, C. S.: 16, 21
Gulf Oil Corp.: 6, 7, 26, 27, 36, 38

Halverson, James T.: 87
Hanna Mining: 45
Hardesty, Howard: 80
Harkness, Stephen V.: 2
Harriman, Edward: 5
Harris, Oren: 50
Henderson, Loy: 37
Hepburn Act: 11
Herrington, Carl: 79
Holding companies: 52–59, 64–65, 123; banks as, 65, 66–67
Hoover, Herbert: 9
Hoover, Herbert, Jr.: 36, 37, 70
How to Challenge Your Local Electric Utility: A Citizen's Guide to the Power Industry: 151
Hughes, Joseph: 133–135
Humble Oil and Refining Co.: 6

Ibn Saud: 24, 31
Ickes, Harold: 23–25, 28, 144
I. G. Farben: 85
Independent Petroleum Association of America: 86
Independents, restricted role of: 7–8, 10, 12, 13, 76, 78, 100, 110
Indiana Natural Gas and Oil Co.: 47
Information, industry control: 86–87, 89, 90; reform, 106–109
Insull, Samuel: 55–56
Interior Department: 28, 72–73, 82, 167
Interlocks, corporate: 48–49, 53, 57–58, 65, 75, 130
Internal Revenue Service, industry-benefiting rulings: 33, 81
Interstate Natural Gas Co.: 48
Iran: xii, 31–32, 37, 70
Iraq: xi, 16–17, 21
Iraq Petroleum Co., Ltd. (IPC): xi, 16–17, 20; Red Line Agreement, 20–22
Israel, U. S. policy to: 30–31
Italy: 70

Jackson, Henry: 89, 101–102, 108, 109, 110
Johnson, Lyndon B.: 50, 81, 89
Joint ventures: 70, 74, 75–77
Josephy, Alvin M., Jr.: 83–84
Judge, Thomas: 152

Kaplan, Richard: 146
Kerr, Robert: 50
King, A. P., Jr.: 86
Kirschner, Edward: 149
Koppers Co.: 57
Krug, J. A.: 28
Kuwait Oil Company: 26

LaFollette, Robert: 9
Land and Water Conservation Fund: 102
Lane, Franklin K.: 9
Lee and Company: 57
Lehigh Coal Co.: 39
Levinson, Jerome: 97
Lewis, John L.: 78
Lifeline Service: 136–141
Lone Star Gas Co.: 49
Love, George: 78
Lurgi Company: 85
Lybian National Oil Co.: 96

McAdoo, William: 9
McCormack, Mike: 118, 119
McDonald, Thomas: 40
McGhee, George: 30–31, 32, 33
McGrannery, James P.: 35
McKay, Douglas: 66
Manning, Richard: 95–98
Mansfield, Mike: 104

Index 223

Massachusetts, resistance in: xii, 138–141
Mattei, Enrico: 70
Mead, Walter: 73–74
Middle East: xi, 19–22, 30–31, 34
Middle South Utilities: 65
Middle West Utilities Co.: 56
Miller, Arnold: 105
Miller, Otto: 77
Mills, Wilbur: 115
Mineral Leasing Act: 82
Mississippi River Fuel Co.: 48, 49, 58
Mitchell, S. Robert: 80
Morgan, John P.: 52–53, 57, 58
Morgan interests: 48–49, 169
Mossadegh, Mohammed, ousted: xii, 31–32, 37, 70
Muscle Shoals case: 59–60

Nader, Ralph: 110, 113
National energy organization, proposed: 109–112, 167–170
National Environmental Policy Act: 102
Nationalism: xii, 31–32, 34–35, 45, 70, 72
National Petroleum Council: 28
National Power and Light Co.: 57
National Recovery Administration (NRA): 10
National Science Foundation: 116, 159
National Security Council: 32, 37
Natural Gas Act: 50
Natural gas industry: 47–51, 86
Natural Gas Pipe Line Co.: 48, 49
Nelson, Gaylord: 107–109
New England Petroleum Corp.: 94–98, 154
Nielands, J. B.: 142–143, 145–146
Nixon, Richard M.: 89, 92–93, 165
North Central Power Study: 83
Northern Plains Resource Council: 85, 103, 104, 151–152
Northern States Power Co.: 61
Nuclear power: 61, 79, 103, 161

Ohio Oil Co.: 14
Oil industry: 6, 17; government and, xi, xii, 6–7, 23–29, 32, 72–74, 79–81, 98; origins, 1–5; vertical integration, xi, 1, 6, 54, 72–75
Olds, Leland: 50
Organization of Petroleum Exporting Countries (OPEC): xii, 72, 160
Outer Continental Shelf: xii, 72, 100, 103, 168
Outer Continental Shelf Act: 29

Pacific Gas and Electric Co.: 63, 142–152
Pacific Oil: 6

Packard, Walter: 148
Panhandle Eastern Pipeline Co.: 49
Participants Agreement: 37–38
Penn, Thomas: 43
Perry, E. Ray: 131
Petrakis, Peter: 146
Petroleum Reserves Corp.: 24–25
Petulla, Joseph: 147, 148, 149
Phelps, Edward: 104
Philadelphia and Reading Coal and Iron Co.: 45
Phillips case: 50–51, 85–86
Pinchot, Gifford: 9
Pipelines: 3, 23, 47; industry control, 10–12, 48–49, 77–78
Pretyman, E. G.: 15
Profits, industry: 5, 100, 124
Project Independence: 103
Prorationing laws: 10, 20, 90
Public Energy Districts (PED), proposed: 163–165, 170
Public Service Co. of Northern Illinois: 55
Public Service Corp. of New Jersey: 58
Pure Oil: 6
Puterman, Martin: 138–141

Quotas, import: 71, 90–91

Railroads, rebates: 2–3, 11
Raker, John: 142
Raker Act: 142–143, 144, 146
Rates, utility: 54, 59, 61, 149; discriminatory, 125, 126, 127, 137, 139–140
Ray, Dixie Lee: 116, 117
Rayburn, Sam: 49
Red Line Agreement (IPC): 20–22
Refining: 12, 91–92
Reform efforts: 101–119; industry data, 106–109; strip-mining, 103–106
Regional energy boards: 165–167
Research industry: 119
Reserves, fuel: 7–8, 8–9, 89
Resistance: xii, 85, 103–104, 122–151
Rhode Island Utilities Commission: 61
Rockefeller, David: 131
Rockefeller, John D.: 1–5, 6, 11
Rockefeller, William: 2, 4
Rockefeller family, power interests: 14, 48–49, 169
Rodberg, Leonard: 162
Rogers, H. H.: 5
Roosevelt, Franklin D.: 23, 24, 25, 49, 59
Royal Dutch Shell: 16, 18, 20, 26, 27
Rural Electric Cooperative System: 66

224 Index

Rural Electrification Administration: 63

San Francisco, resistance in: 142–152
San Francisco Examiner: 144
Saudi Arabia: xii, 21, 24, 32, 70
Shale oil: 18, 103, 168
Sieberling, John: 104–105
Simon, William: 107, 109
Sinclair, Harry: 9
Skinner, Scott: 136
Socony Mobil: 26, 27
Socony-Vacuum Oil Co.: 7, 13, 14, 20, 21, 36, 38
Solar energy: 115–119, 161, 171–173
Southeast Asia: xii, 72
Southern California Edison Co.: 63
Southern Co.: 122, 128, 130
Southern Electric Generating Co. (SEGCO): 130–131
Soviet Union: 30–31, 71
Sratchona, Lord: 15
Staggers, Harley: 109
Standard Oil Co.: 1–5, 6, 16, 17
Standard Oil Co., Calif.: 1, 6, 14, 21, 24, 26, 36, 38, 49, 76, 94–98
Standard Oil Co., Ind.: 14
Standard Oil Co., N. J. (Exxon): 6, 7, 14, 19–20, 26, 27, 36, 38, 48, 49, 79, 85
Standard Pacific gas pipeline: 49
Standard Power and Light: 58
State Department, industry ally: 1, 19, 26–27, 30, 34, 98
Stevenson, Adlai, III: 109
Stiles, Stan: 93
Stillman, James: 4–5
Stocking, George W.: 32
Stork, Joseph: 71
Strip-mining: 44, 64, 81, 103–106
Submerged Lands Act: 29
Sullivan, Jean: 146
Supreme Court: 10, 50–51, 61, 85, 145
Swidler, Joseph: 86

Symington, James: 118

Tax policy: coal, 45–46; depletion, 9, 113–114; drilling costs, 114; foreign tax credits, 32–33, 114–115; reform effort, 112–115
Teapot Dome: 8–9
Temporary National Economic Committee (TNEC): 7, 8, 10, 48, 58, 59
Tennessee Valley Authority (TVA): 59, 61, 62, 63, 78–79, 165–166
Texas Co.: 6, 7, 24, 26, 36, 38
THUMS: 76
Tidelands oil issue: 28–29
Tidewater Oil Co.: 3–4
Transportation, industry control: 3–4, 10–12, 39, 40–41, 42–43, 48–49, 169
Truman, Harry S: xii, 28–29, 30, 35, 50
Truppner, William C.: 92
Turkish Petroleum Co.: 20
Turner, Donald: 81

Union Oil: 6
Union Pacific Railroad: 5
United Corp.: 57
United Founders Corp.: 57
United Gas Corp.: 49, 58
United Mine Workers (UMWA): 44, 45, 78, 105–106, 133, 150
Utility Holding Company Act: 59, 60, 131, 133

Venezuela: 31
Vermont, resistance in: xiii, 136–138
Vogtle, Alvin: 130

Walsh, Louis J.: 12
Water resources: 81–82, 88, 106
Webb, Lee: 136–138, 163
Whisnant, David: 163–164
White, David: 18
White, Lee: 110
Willis, Glenn: 92
Wilson, John: 74–75
World War II: 21, 23–29